D0930538

THE WETLANDS

Look for these and other books in the
Lucent Endangered Animals and Habitats Series:

The Amazon Rain Forest
The Bald Eagle
Birds of Prey
The Bear
Coral Reefs
The Elephant
The Giant Panda
The Gorilla
The Manatee
The Oceans
The Orangutan
The Rhinoceros
Seals and Sea Lions
The Shark
The Tiger
The Whale
The Wolf

Other related titles in the Lucent Overview Series:

Acid Rain
Endangered Species
Energy Alternatives
Garbage
The Greenhouse Effect
Hazardous Waste
Ocean Pollution
Oil Spills
Ozone
Pesticides
Population
Rainforests
Recycling
Saving the American Wilderness
Vanishing Wetlands
Zoos

THE WETLANDS

BY DANIEL KRIESBERG

San Diego • Detroit • New York • San Francisco • Cleveland • New Haven, Conn. • Waterville, Maine • London • Munich

Dedication
To Zack and Scott and all the wetlands we will explore together. Also, a special thanks to all the people who talked with me about wetlands, especially Dr. Steve Rice, Dan Ramer, Cassandra Harper, and my editor Lauri Friedman.

For more information, contact
Lucent Books
27500 Drake Rd.
Farmington Hills, MI 48331-3535
Or you can visit our Internet site at http://www.gale.com

LIBRARY OF CONGRESS CATALOGING-IN-PUBLICATION DATA

Kriesberg, Daniel A.
Wetlands / by Daniel Kriesberg
p. cm. — (Endangered animals and habitats)
Includes bibliographical references.
Summary: Explains what a wetland is, why it is important, threats facing this ecosystem, and efforts to protect wetlands.
ISBN 1-56006-923-6 (hardback : alk. paper)
1. Wetlands—Juvenile literature. [1. Wetlands. 2. Conservation of natural resources.]
I. Title. II. Endanered animals and habitats.
QH87.3 .K75 2003
333.91'8—dc21

2002013938

Printed in the United States of America

Contents

INTRODUCTION 6
Changing Wetlands

CHAPTER ONE 9
What Is a Wetland?

CHAPTER TWO 26
The Benefits of Wetlands

CHAPTER THREE 43
Wetland Destruction

CHAPTER FOUR 57
Wetland Degradation

CHAPTER FIVE 73
Protecting Wetlands

NOTES 89
GLOSSARY 91
ORGANIZATIONS TO CONTACT 93
SUGGESTIONS FOR FURTHER READING 95
WORKS CONSULTED 97
INDEX 103
PICTURE CREDITS 111
ABOUT THE AUTHOR 112

Introduction

Changing Wetlands

WETLANDS ARE USUALLY not considered tourist attractions or popular vacation spots. The images most people have of swamps, marshes, and bogs are of muck, stink, and scary creatures. In fact, until recently, most wetlands were often considered wastelands.

Despite this reputation, wetlands are a very valuable resource. The first civilizations in human history were dependent on wetlands for their survival. The Mesopotamians, the Egyptians, and other ancient people lived in or near wetlands. They took advantage of the wetlands' water, plants, and animal life for their food, water, and shelter. Today, people are still dependent on wetlands for many everyday resources. We even depend on ancient wetlands for the fossil fuel we use for gas and oil, and to generate electricity.

There are a wide variety of wetlands. They are found on every continent and vary in size from a few acres to millions of acres. Wetlands can be found on mountains, in valleys, in jungles, and north of the Arctic Circle. They are created in different ways depending on the location and type of wetland. Most of the wetlands in Alaska, Canada, and the northern third of the United States were created by glaciers. Other wetlands were created by the actions of lakes and rivers. Geologic forces created the conditions for wetlands in the Florida

Everglades, where limestone located underneath the ground holds water. Other natural forces such as beavers, storms, and droughts can cause new wetlands to be created and existing wetlands to be destroyed. Sometimes wetlands even change from one type to another.

Wetlands offer diverse habitats to countless species of mammals, birds, fish, reptiles, amphibians, insects, invertebrates, and plant life.

Because of wetlands' enormous diversity, countless animals make their homes there. Mammals, birds, fish, reptiles, amphibians, and vast numbers of invertebrates spend all or part of their lives in the world's wetlands. Wetlands are considered biological supermarkets because of their extensive food chains and biodiversity. An astounding variety of plants can also be found in wetlands.

Unfortunately, wetlands, like many ecosystems around the world, are threatened by a variety of human actions. The changes to wetlands caused by people are different from the changes resulting from natural forces. Human-caused changes happen too quickly for most animals and plants to adapt. People have drained wetlands of their water or filled them in to make more room for farming and development. As technology has improved and the demand for land and resources has increased, huge areas of wetlands have been altered. As of 2000, approximately half of the world's original wetlands have been destroyed. And the remaining wetlands face a wide variety of threats.

Although these threats continue to alter wetlands, two factors have changed the way that many people view these natural resources. First, it is now clear that there is not an unlimited amount of wetlands, and we must work to protect the ones that remain. Second, there has been a greater understanding of the benefits that wetlands provide for all life on earth. Recognizing the importance of wetlands and the threats to their continued existence has led to efforts to preserve and restore wetlands around the world.

1

What Is a Wetland?

WETLANDS ARE ONE of the most interesting ecosystems on the planet yet one of the hardest to define. Wetlands are not aquatic ecosystems such as lakes or ponds. They are also not terrestrial (land) ecosystems like forests or deserts. Instead, wetlands fall somewhere in between. The words *wet lands* explain what makes a wetland a separate ecosystem. Wetlands are lands where water defines the type of soil and the types of plant and animal communities living there.

There are many different definitions of what a wetland actually is. The most comprehensive definition of *wetlands* is the one used by the U.S. Fish and Wildlife Service: "Wetlands are lands transitional between terrestrial and aquatic systems where the water table [the top of an underground bed of soil or rock that is saturated with water] is usually at or near the surface or the land is covered by shallow water."[1]

Agreeing on a definition of a wetland is an important task for scientists and government officials. What makes this task difficult is the fact that wetlands vary a great deal. Determining what constitutes a wetland dictates countless decisions on how the land can be used, how and where conservation money is spent, and what environmental laws are enforced.

Some wetlands are easy to identify, but others present some difficulty because they do not always have standing water. The three indicators scientists look for to define an area as a wetland are water, soil conditions, and the plant species growing at the site.

Water in wetlands

All wetlands contain water during at least some part of the year. Wetlands may be covered in water year-round or be wet only during certain seasons. As long as the land is flooded for a certain number of days in a row during the growing season, the area is considered a wetland. (A growing season begins when the first buds appear and ends with the first frost.)

Wetlands get their water from a variety of sources. Some wetlands are located in places with a high water table, meaning that underground water is near the surface. The water in a wetland may also come from rain, snow, and other forms of precipitation. Other wetlands contain surface water from rivers, lakes, oceans, or other bodies of water. The water in wetlands can be salt water, freshwater, or brackish (a mixture of salt water and freshwater). Wetland water is usually no more than six feet deep, which is shallow enough for sunlight to reach the bottom, thereby allowing plants to grow in the water.

Giant cypress trees flourish in this southeastern U.S. wetland despite very wet conditions.

Since water is not always present in a wetland, scientists look for clues to determine if a site is covered in water on a temporary basis. Water will often leave behind stains on rocks, bark, or leaves. Sometimes debris gets left behind from past floods. Evidence of water is not limited to the surface. By studying the soil, scientists can identify whether an area is truly a wetland.

Wetland soil

The soil in a wetland is called hydric soil (*hydric* means "contains lots of water"). Hydric soil is saturated with water during all or part of the growing season, giving it characteristics that differentiate it from soil in other places. Hydric soil is periodically deficient in oxygen, which can make it difficult at times for plants to get the oxygen they need to grow.

The lack of oxygen in hydric soil also slows the ability of bacteria to decompose organic materials. (Organic material is made up of dead and decaying plants and animals.) Therefore, another characteristic of some soil in wetlands is a high level of organic material at its surface. This organically rich soil greatly affects the types of plants that grow in wetlands.

Wetland plants

Though it might seem like plentiful water would not be a problem for plants, it does present difficulties. Too much water restricts the oxygen in the soil, and plants need oxygen for respiration (the cellular process that plants and animals use to obtain energy for growth). Because each plant cell carries out the process of respiration, plants need oxygen above and below the ground for respiration to occur. Thus, only certain plants can live in a place with very little oxygen.

The plants that live in wetlands are called hydrophytes (from the Greek words for "water plant"). Hydrophytes have many special adaptations for living in an area that is dominated by water and low in oxygen. Wetland plants often have roots near the surface of the soil to obtain more oxygen. When oxygen does become available, many wetland plants use adventitious roots to find oxygen; these are roots that can grow and spread quickly to oxygen-rich places in the soil. Other plants have hollow stems that allow air from the leaves to go to the roots.

There are three major groups of wetland plants, which are categorized by where in the wetland they grow. Emergents are plants that grow partly in the water and partly out of the water. Cattails, sedges, and tule are examples of emergents. In deeper water, floating aquatics dominate. These are plants with roots in the mud and leaves that float on the surface. The water lily is an example of a floating aquatic. The last group is the submergents, which grow completely

Floating aquatic plants, like this species of water lily, collect sunlight on the surface while maintaining roots in the soil below.

underwater. Elodea, a common plant used in many home aquariums, is an example of a submergent. Submergents have roots in the soil and flexible stems.

Defining wetland types

It is not always clear exactly when a wetland ends and a pond, lake, or ocean begins, or where a forest, jungle, or tundra ends and a wetland begins. The natural world is not organized with clear boundaries. Wetlands cannot be separated by fixed markings the way a fence separates two pieces of property. The edges of different types of wetlands often blend into one another as the conditions vary. Therefore, a given wetland may have areas that contain many different features.

Wetlands can be classified in several ways. Some scientists use the water source as a criterion for classification. Other ways include grouping wetlands by their location, by whether the wetland is coastal or inland, or by whether it contains freshwater or salt water. The major types of wetlands—bogs, marshes, and swamps—exhibit an enormous amount of diversity.

Bogs (peatlands)

Bogs are one type of wetland found across areas of the Northeast, the north-central United States, Canada, Siberia, and other parts of the Northern Hemisphere. Bogs are located in what were once deep glacial lakes. They formed over time as these lakes became filled with peat (layers of dead, partially decomposed plants).

The peat in bogs, or peatlands, lies over the water and thickens as new layers are continually added. This creates a floating mat of vegetation that eventually separates plants from the groundwater and soil. The live plants grow on the peat, not on the water and soil. Because it is cut off from the groundwater, life in the bog is dependent on rain and snow for its water supply. The lack of water movement from below fails to carry nitrogen, phosphorus, and other nutrients to the plants. Instead, the nutrients are trapped in the peat and not available for growing plants.

Bog Formation

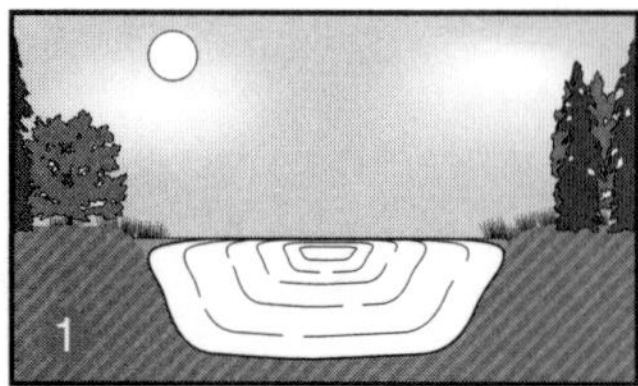

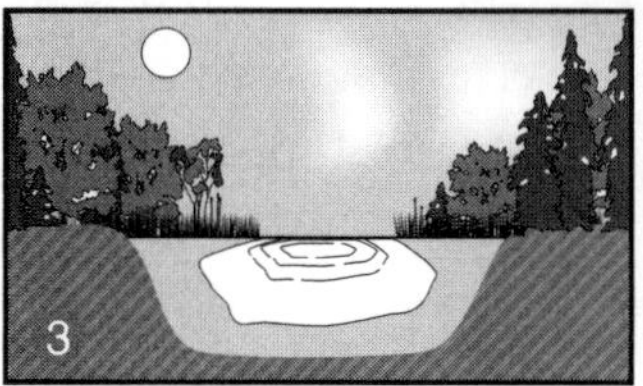

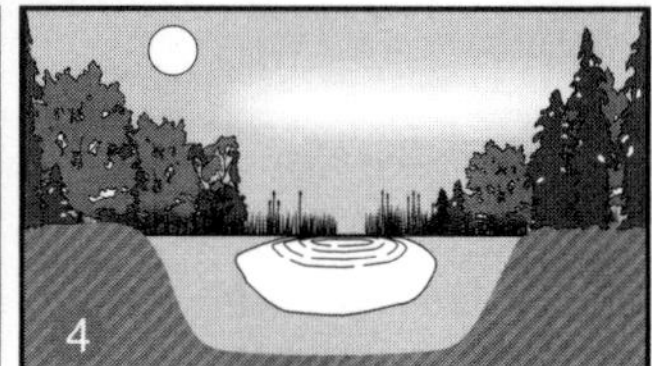

1. A vegetative mat, primarily consisting of sphagnum moss, begins to grow out from the shore along the surface of an area of open water.

2-4. Peat continues to accumulate and the floating mat grows larger. The water develops a mucky bottom.

5. The open water becomes completely cut off, and the floating mat consolidates.

Few plants can grow in these harsh conditions. A three-hundred-year-old tamarack tree may grow only four feet tall. However, one plant that thrives in these conditions is sphagnum moss. This live moss grows on top of the layers of peat, forming a living skin on the surface of decaying plants. Sphagnum moss can absorb up to one hundred times its weight in water, enabling it to retain water for long periods of time. Sphagnum moss tends to dominate the bog. According to Steve Rice, a botanist at Union College, "Sphagnum has an enormous capacity for scavenging nutrients. . . . The consequence of this is the bog becomes even more suitable for sphagnum and less so for other plants."[2]

Since there are so few obtainable nutrients in bogs, other plants that survive there have some interesting ways for acquiring the nutrients they need. One method is to capture and ingest insects. Sundews and pitcher plants are two examples of insect-eating plants that can be found in bogs. Sundews catch insects with sticky leaf pads that curl up and digest their prey. The leaves of the pitcher plant form a vaselike shape; insects fall in, drown, and are digested in the water that collects in the bottom.

Compared with other kinds of wetlands, there is less plant variety in a bog. As a result, there is also less animal life. Moose, bears, otters, and other animals may visit bogs in their search for food, but they do not stay. Bogs are also an important stopping place for some migrating birds. One animal that does spend its life in a bog is the bog lemming. Bog lemmings look like large mice with very short tails. They live in colonies and create complex tunnel systems through the bog. They eat leaves, stems, and seeds. It is believed that bog lemmings' tunnels play an important role in recycling some of the nutrients in a bog.

Fens

Like bogs, fens accumulate layers of peat, have glacier origins, and are found in low points on the landscape, usually in depressions on mountains and mountainsides and occasionally in lowlands. It is the source of water that is the difference between a fen and a bog. Unlike bogs, fens receive water from underground sources as well as from precipitation. Fens are the only type of wetland primarily dependent on groundwater sources.

Because fens receive nutrients from groundwater, they are more productive than bogs and are able to support a greater variety of plant and animal life. Fens look somewhat like meadows. Instead of sphagnum moss, fens contain grasslike plants called sedges and other nonwoody plants. Fens with low amounts of nutrients occasionally turn into bogs as more peat accumulates.

Bog People

One of the amazing things about bogs is that dead plants there decompose very slowly, if at all. Because decomposition in bogs is limited, dead things are preserved remarkably well. For example, in 1950, two farmers were cutting peat in a bog near Tollund, Denmark. While digging, they found a body of a man. The body was so well preserved that at first they thought they had discovered a recent murder victim, and they notified the police. Upon investigation, it was discovered that the man had been killed over two thousand years ago. The man had been put to death with a rope around his neck and placed in the bog, where his remains became mummified.

The amazing preservation of the Tollund man allowed scientists to figure out that his last meal was a soup made of grain and seeds. The type of grain they found in his stomach helped the scientists determine when the man had lived. The food also helped the scientists approximate the time of year he was killed. Carbon-14 dating on a wooden tool found nearby the body put the date of his death around 220 B.C.

Over the years, about two thousand bodies or parts of bodies have been found in the bogs of northern Europe. The bogs preserved the remains so well that in one case a man's fingerprints were still clear. Police compared them to those of people today and found that they were similar to the fingerprints of many people living in Denmark.

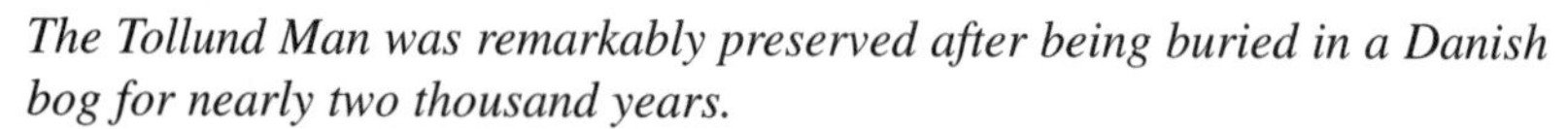

The Tollund Man was remarkably preserved after being buried in a Danish bog for nearly two thousand years.

Cattails, emergent wetland plants, are critical links in the freshwater marsh ecosystem.

Freshwater marshes

Another type of wetland is the marsh. Marshes are formed in depressions along rivers, streams, ponds, and lakes. They get their water from streams, rivers, snowmelt, flooding, or groundwater. Marshes are characterized by a year-round supply of water and generally high nutrient levels. Their mucky soil holds enough nutrients for a wide variety of plants to grow in, including reeds, sedges, grasses, and other hydrophyte plants. There is too much water in marshes for trees to grow; only on hummocks (mounds that rise slightly above the water) will trees grow.

Depending on the marsh's location, the dominant plant found there varies. The tule marshes in California and Oregon consist mainly of bulrush. These plants grow up to four feet tall and are topped by grassy tufts. Saw grass dominates the huge expanse of marsh that makes up the Everglades in Florida. Cattails are common in marshes in the northeastern United States and Canada. Cattails protrude from the marsh with brown, hot dog-like flowers that make them easy to identify.

Cattails are an important source of food and shelter for animals in a marsh. Native Americans knew the importance of cattails and used every part for food. Even today, cattails have a variety of uses. Adhesives are made from their stems, ethyl alcohol from the flowers, and rayon from the pulp.

The abundant plant life in marshes allows a wide variety of animals to find sustenance there. Ducks, muskrats, mink, snapping turtles, and numerous amphibians are just a few of the animals that find food, water, and shelter in freshwater marshes.

Prairie potholes (seasonal marshes)

A variation on the marsh is the prairie pothole, found in the upper midwestern United States and Canadian prairies. Prairie potholes are one of the several kinds of seasonal wetlands. They were formed during the last ice age. As the glaciers melted back to the Arctic, chunks of ice broke off and became buried in sediment. Eventually, the ice chunks melted and forced the land to slump downward, causing a depression. In the spring, these low-lying areas fill with water from snowmelt and rain. By the fall, the potholes have shrunk greatly in size or dried up completely (some stay dry for several years). By the following spring, when enough rain and snow have fallen, the cycle begins again, transforming parts of the prairie into a marsh.

An estimated 50 to 75 percent of all ducks, geese, and swans in North America are hatched in prairie potholes. Not only are the prairie potholes critical nesting habitats, but they also provide food, water, and shelter for waterbirds during their migration.

Salt marshes

Yet another type of marsh is the salt marsh. Salt marshes form in shallow, calm waters along many coastlines in the Northern Hemisphere. Salt marshes also form in the calmer waters between barrier islands and the mainland and in estuaries (places where salt water and freshwater mix). Salt marshes are influenced greatly by wave action, tides, and the salinity (saltiness) of the water.

Wetland Names

Around the world, there are many interesting local names to describe the wide variety of wetland ecosystems. Bogs, for example, are sometimes called peatlands because they accumulate large quantities of peat. In Britain, bogs are often called moors. The word *moor* comes from "dead" or barren land. There are high moors and low moors depending on the elevation. In other parts of Europe, bogs are often called mires. The Danish and Swedish word for bog is *mose*. In Canada and Alaska, large areas covered with bogs are called muskegs. The peatlands of southwestern New Zealand are called pakihis. In the southeastern United States, there is a unique type of peatland called pocosin where shrubs and evergreens grow. The name *pocosin* comes from the Algonquin Indian word for "swamp on a hill."

Swamps and marshes also have a variety of different names throughout the world. In Australia, there are swamps called billabongs, which are streamside wetlands that are periodically flooded. In the United States, bottomlands are a name used for swamps in low-lying lands along rivers and streams that are flooded on occasion. In Britain, carrs are shrub swamps where alders and willows grow. Cattail marshes in Britain are called reed-mace swamps. A raupo swamp is the name of a cattail swamp in New Zealand. Playas are seasonal marshes in the southwestern United States that are similar to prairie potholes.

Because of the mixture of salt water and freshwater that passes through them, salt marshes have many nutrients. The tides mix the water and nutrients, creating a place with high productivity (the rate that plants produce energy) and biodiversity. Salt marsh plants require special adaptations to survive, because the salt water makes it hard for plants to take in water and nutrients. Twice a day, high and low tides first submerge plants in water for part of the day and then leave them dry. This makes the plants subject to both aquatic and desertlike conditions in the same place.

By far, the most dominant species of plant in the salt marsh is spartina, which is also known as cord grass or salt marsh grass. It is the only plant that can live both in and out of the water daily. Spartina is a halophyte (which means "salt-loving" plant). Other plants cannot survive in salt water because the water in their cells has less salt than the soil and will flow out of the plant through the process of osmosis. Spartina, however, adapts to the high salinity by bringing salt into its roots. With the higher level of salt in the roots, water will flow into the spartina's cells. The excess salt is then excreted through glands. Spartina also has underground stems called rhizomes that help to bind the marsh's mud together, creating habitat for mussels, crabs, and other animals.

One animal that takes advantage of the salt marsh is the diamondback terrapin. These are the only turtles whose habitat is limited to the brackish waters of a salt marsh. According to Matthew Draud, a marine ecologist who studies terrapins in the Long Island Sound, "Terrapins are important consumers in the estuarine food webs. Their diet includes mainly marine mollusks, such as snail, clams, and mussels. A favorite delicacy for the diamondback terrapin is the fiddler crab."[3]

Early in the twentieth century, diamondback terrapins were considered a delicacy themselves and were used for soup. Fortunately, protective legislation has helped to restore the diamondback terrapin to a healthy population.

Swamps

The third major type of wetland is a swamp. Swamps are the most abundant type, making up more than half of all wetlands. Swamps are dominated by trees. They vary from the brilliant red maple swamps of the northeastern United States to the stately cypress swamps of the Southeast. There are even rain-forest swamps, where river dolphins swim among giant trees.

Swamps are often formed when a river nears the low elevations of a coastline, slows down, and creates standing water. Swamps are also found along lakes and other places of

The Largest Wetlands

Wetlands come in all sizes, and two of the biggest are the Gran Pantanal and the Okavango Delta. The Gran Pantanal wetlands, one of the largest and most unique wetlands in the world, are found along the border of Brazil and Paraguay in the center of South America. These wetlands cover more than 34 million acres, which is four times the size of the Florida Everglades. The Gran Pantanal has a wet and a dry season. During the wet season, wildlife is spread out over the entire area. During the dry season, when water becomes scarce, the animals gather around the few sources of water. The Gran Pantanal is not the best known place in the world, but it is famous for its bird life. There are about seven hundred species of birds living there, including the jabiru, the largest flying bird in the Western Hemisphere. The largest rodent in the world, the capybara, and a great deal of other wildlife also live in the Gran Pantanal.

Another extensive wetland is the Okavango Delta in Africa. This inland delta is formed at the point where the Okavango River and the sands of the Kalahari Desert meet in Botswana. It covers more than 16 million acres. The Okavango Delta is a seasonal wetland that creates a web of lagoons, channels, and islands, which become home to crocodiles, elephants, buffalo, and more than four hundred bird species. People living in the area also use the Okavango Delta for food, water, and shelter. The wonders of the Okavango Delta have made it an increasingly popular tourist destination.

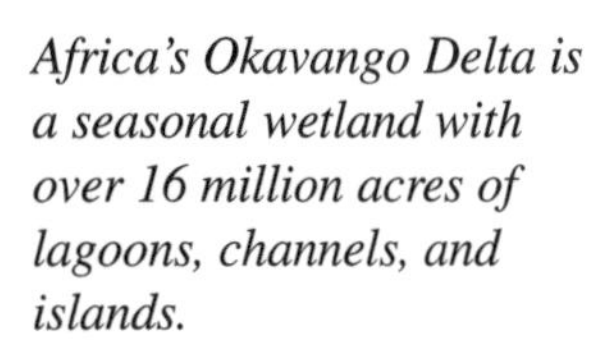

Africa's Okavango Delta is a seasonal wetland with over 16 million acres of lagoons, channels, and islands.

low elevation where water collects. Trees can survive in swamps because they are covered in water only seasonally. Trees also prosper because of the high nutrient content of the soil, which is enhanced when floods carry in and deposit sediment.

Trees that live in swamps usually have shallow roots. Being close to the surface allows the roots to obtain more oxygen, instead of growing deep into the oxygen-deficient soil. Shallow roots, however, provide less support, and many swamp trees are in danger of falling over. Some swamp trees have adapted by developing buttressed tree trunks for extra support. These extra thick trunks keep the trees from toppling.

The classic example of this tree-trunk adaptation is the bald cypress tree of the southeastern United States. Bald cypress trees, relatives of the redwoods, are tall and majestic, with a flat-topped, irregular shape. Nicknamed "wood eternal," bald cypresses are extremely resistant to decay. Because of these features, bald cypresses were (and sometimes still are) heavily lumbered and used for building docks, boats, bridges, and other items that come in contact with water.

Natural resistance to wood-rotting fungi helps bald cypress trees survive swampy environments.

Vernal pools

There is one type of swamp that, although small, is a critical wildlife habitat: vernal pools. Vernal pools are found in small shallow depressions in forests. They are created when snowmelts and spring rains fill these depressions, which hold water through winter and spring but usually dry out by the end of summer. Because of the temporary nature of vernal pools, fish cannot live in these wetlands. Thus, with no fish to prey on their eggs, tadpoles, salamanders, and other aquatic larvae have a better chance to survive. This makes the pools important breeding grounds for salamanders, frogs, and toads. After the first spring rains, these amphibians come out of hibernation and the pools come alive with mating and egg laying. By the time the vernal pools dry up, the amphibians have developed and moved on to their lives on land.

Many insects also use the pools to lay their eggs. The larvae complete their development and fly off as the pool dries up. Other animals in vernal pools, such as fairy shrimp, bury themselves in the mud to wait out the dry season.

Mangrove swamps

Another type of swamp is the mangrove swamp. Mangrove swamps are saltwater swamps found in warm, tropical waters that have average temperatures of 24 degrees Celsius or higher. They grow in shallow, protected areas along the coastline. Mangroves are characterized by tons of mud, making it virtually impossible to walk through them. Out of the mud comes a tangle of roots propping up strange-looking trees called mangrove trees.

Mangrove trees have to contend with many of the same problems that plants in the salt marsh face—namely, salt and mud. There are three ways mangrove trees rid themselves of salt. Some filter it out at the root level before the salt even gets in. Others have glands that secrete the salt out onto the leaves. A few species concentrate the salt in the bark or older leaves; when these parts of the plants fall off, they take the salt with them.

Tangled mangrove pneumatophores provide a backdrop for a lone Louisiana heron scanning for food.

Mangrove trees have adapted to the lack of support and oxygen in the muddy habitat by developing extensive root systems. Mangrove species that live in more saturated soils have parts of their roots raised above the soil to take in oxygen. These structures, called pneumatophores, stick out of the mud like fingers. Another adaptation to the muddy soil is seeds that sprout while still attached to the adult plant; when the seeds fall into the unstable mud, they have a head start in gaining a secure foothold.

Mangrove swamps are home to many kinds of crabs, snails, snakes, raccoons, birds, crocodiles, manatees, and more, who take advantage of the food and shelter that mangroves provide. The fascinating habitat of mangrove swamps has been the subject of a great deal of research, not only because of the swamps' worldwide range but because of their many unique features.

Wetlands are found on every continent except Antarctica. They can be found high in the mountains or along the seashore. There are wetlands in hot, steamy jungles, on frozen Arctic tundras, and even in suburban and urban areas. They range in size from a few square feet to thousands of square miles. With such a wide variety of wetlands and with wetlands in so many places, it is no surprise that wetlands are home to a diversity of plants and animals. However, the value of wetlands does not end with the plants and animals they support. There are also many ways wetlands benefit humans.

2

The Benefits of Wetlands

THE VALUES OF wetlands are only recently becoming known and understood. Work by scientists begun in the 1960s and 1970s is now revealing the important role that wetlands play in the environment and the many ways that humans depend on wetlands for flood control, food, water, recreation, ecological stability, and even for the air we breathe.

The benefits derived from wetlands come from their functions. Scientists use the word *function* to describe what wetlands and other ecosystems do. Functions are the physical, chemical, and biological processes that happen in and make up an ecosystem. Different ecosystems have different functions, and even different wetlands have different functions. Some wetlands perform certain functions more effectively than others, and some work together to perform a function. For example, a group of wetlands can act as a chain along a migration route for shorebirds that need food and resting places in order to complete their trip.

Many of wetlands' functions are important to humans. Wetlands play a critical role in the many ecological cycles that provide humans with food, water, and air. They also protect people from pollution and other human-made problems. Wetlands protect people from natural forces on the planet as well, such as flooding.

Flood control

Flooding can cause devastating damage to property and claim lives. Wetlands play an important role in flood control. Approximately 134 million acres of land in the United States have severe flooding problems. Most of this is farmland, but about 3 million acres are in urban areas. Each year, floods cause about $3 billion in damage. The severity of floods has increased over the past few decades as more natural areas have become developed. Asphalt, concrete, and pavement prevent water from soaking into the ground and instead deposit it directly into waterways. The development, therefore, is responsible for increased runoff into streams, creeks, rivers, and lakes. The large water load raises the flow of rivers and streams during times of heavy rain or snowmelt.

Wetlands can ease and even prevent the effects of a flood by acting as a giant sponge, absorbing and storing water. The water is absorbed into the wetland soil and is released slowly into lakes, rivers, and streams. In addition, the leaves and roots of wetland plants slow the water down as it moves through the wetlands. This allows the water to run off at a slower pace, giving floodwaters more time to recede. Wetlands in and around urban areas are particularly important because of the increased runoff in cities and the potential for greater damage.

Wetlands are less expensive and often more effective for flood control than building levees (embankments built along rivers to prevent overflow). In most cases, man-made flood-control measures divert floodwaters to somewhere else downstream and simply move the flood to a new location. Along the Charles River in Boston, the Army Corps of Engineers realized that it was more cost-effective to buy the wetlands along the river and preserve them instead of building flood-control structures to protect the city. If 40 percent of the wetlands along the Charles River and its tributaries were destroyed, it was estimated that the flood damage would be $3 million a year. If all the wetlands were destroyed, the damage from a flood would cost $17 million. To prevent this, the Army Corps of Engineers decided to buy the floodplain wetlands. It has now acquired eighty-five hundred acres of wetlands for flood prevention.

Another area where wetlands are being considered as a flood-control measure is New Orleans, which lies, on average, eight feet below sea level. New Orleans is squeezed between Lake Pontchartrain and the Mississippi River. A severe hurricane in the Gulf of Mexico or a flood on the Mississippi River could cause devastating flooding, putting most of the city under twenty feet of water. To add to the threat, the protective wetlands around New Orleans have been and are being destroyed.

Several plans to protect New Orleans from possible flooding are being discussed, including an idea for a wall that would surround one section of the city. However, according to an article in the *New York Times* on April 30, 2002, "Perhaps the surest protection is building up the coastal marshes that lie between New Orleans and the sea and that have been eroding at high rates."[4]

Preserving wetlands along the Charles River in Boston prevents flooding and conserves natural habitats.

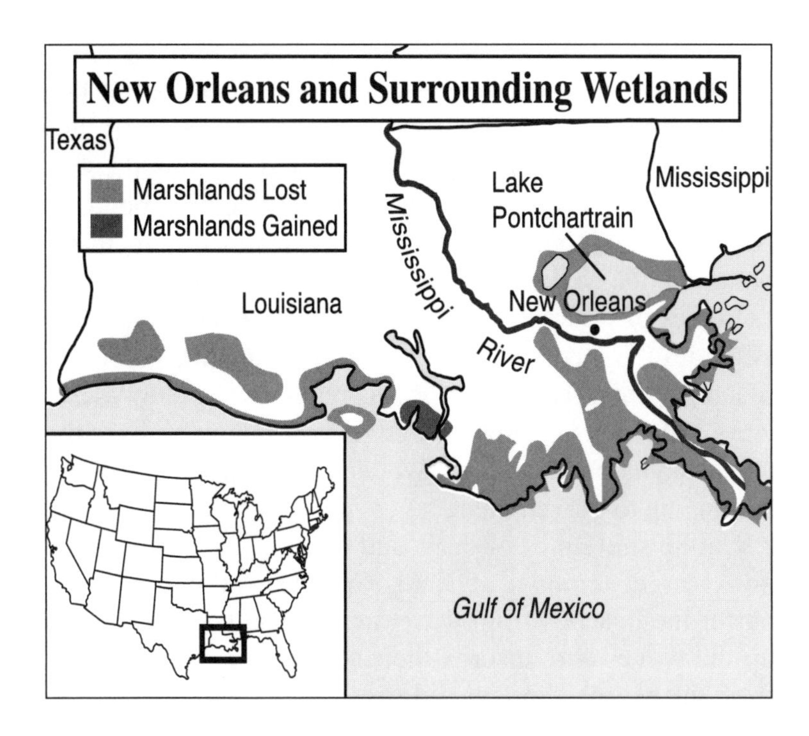

The long-term solution is to slow the loss of the wetlands around New Orleans. One idea is to use gates on the Mississippi River to pour water over the wetlands. This water contains fine bits of soil called sediment. This sediment will help raise or at least maintain the height of the wetlands, which will be better able to hold back floodwaters. Hopefully, the people of New Orleans have not waited too long to realize the importance of wetlands.

Erosion

In addition to floods, water can pose other problems as well. Erosion, the washing away of soil and land by water, threatens farmlands, homes, buildings, and beaches. In New York State, one to four and a half feet of shoreline are lost per year to erosion. In other areas, valuable soil is washed from farms, making it more difficult to grow crops. On beaches and shorelines, waves wash away the land from under houses and buildings. Sometimes so much soil is eroded away that buildings collapse.

Wetlands, however, can prevent erosion by slowing down the rate that the water flows. Slower-moving water has less force and therefore causes less erosion. Wetland plants also prevent erosion by holding in the soil and trapping the sediment, which further prevents land from being eroded. In addition, when the water slows down, sediment falls to the bottom of the wetland instead of being washed away.

Bodies of water downstream from the wetlands also benefit from the wetlands' ability to trap sediment. When too much sediment is washed downstream, it can clog waterways and require expensive dredging operations to make the water navigable for boats. Wetlands help prevent this by holding on to the sediment.

On the shorelines of lakes and oceans, waves can wash away large amounts of sand. Coastal wetlands help to buffer the beaches from wave action. The wetland breaks up the waves and absorbs their effect. This keeps waves from hitting the shore at full force, which would increase erosion. Even in extreme cases such as during a storm or a hurricane, coastal wetlands can lessen the effects of the waves. Reducing the force of the waves, and thus the amount of erosion, can prevent flooding or buildings from collapsing, thus potentially saving lives. The wetlands save money as well. Lessening the waves' impact leads to permanent damage to homes and other man-made structures. Thus, there is less need for insurance companies to make payments or for the federal government to supply emergency funds, thereby saving money for taxpayers.

Water quality

Wetlands also play a role in keeping water clean and pure, and with 40 percent of America's freshwater unusable because of pollution, this is a valuable function. The soil, plants, microorganisms, and even some animals in a wetland help to filter pesticides, herbicides, factory wastes, heavy metals, and other pollutants out of water.

Wetlands cleanse water of pollutants through several processes. Wetlands slow the water flow, which allows time for the pollutants to fall down through the water and

A Salt Marsh Classroom

Salt marshes are great places to get muddy and learn about wetland ecosystems. A walk through a salt marsh is guaranteed to bring people in close contact with amazing plants and interesting wildlife. The environmental education programs at the Waterfront Center in Oyster Bay, New York, allow people to do just that. School groups of all ages who come to the center are led through marsh ecology programs by local naturalists.

A typical salt marsh program takes the students right in to the wetlands. The best time to go is during low tide, when there is more area to explore. Armed with nets, shovels, trays, magnifying glasses, and identification books, the students discover the wonders of a salt marsh. They often encounter animals such as snails, fiddler crabs, hermit crabs, mussels, and all types of marine worms. One of the more exciting animals found there is a soft-shell clam. These clams defend themselves by squirting out a stream of water from their siphons.

The salt marsh exploration can teach people about their community. Visitors leave feeling better connected to the marsh and with an understanding of their role in protecting the marsh and the marsh's role in protecting them. In an interview, Robert Crafa, director of the Waterfront Center, said, "Wetlands are a magnificent educational resource that challenge students to use all their senses to discover the importance of these areas from habitat value to flood protection."

sink into the soil. Pollutants are also removed from the water through plant growth. When wetland plants grow, they take up a lot of minerals into their roots. Pollutants get mixed up with these minerals and are therefore taken out of the water and put into the plant. When the plants die, the pollutants become buried in the soil. Wetlands with large amounts of organic peat act as permanent burial sites for many harmful chemicals. Because the dead plants decompose at such a slow rate, the pollutants remain out of the water. In these cases, the pollutants no longer come in contact

Some wetlands, like this Louisiana marsh, enhance environmental wastewater facilities by extracting pollutants and chemicals.

with the animals in the wetland and are not as likely to wash downstream into lakes and other bodies of water.

Wetlands are so good at cleaning water that they serve as natural water-treatment plants. Dan Ramer, a sanitary engineer for Oneida, New York, says, "Wetlands are now an accepted method of wastewater treatment. Treatment wetlands combining natural and constructed features are used to enhance a wastewater treatment facility's ability to meet environmental guidelines."[5] Because wetlands can handle so many different pollutants, they can be used to clean domestic wastewater, mining drainage, storm water, landfills, and more.

Many cities rely on wetlands to help create clean water. For example, New York City depends on reservoirs in the Catskill Mountains and east of the Hudson River for its drinking water. By 1989, when this water became increasingly threatened

Night Life in a Vernal Pool

An interesting animal that depends on vernal pools for breeding is the tiger salamander. These amphibians belong to a group called mole salamanders. The name comes from the fact that mole salamanders spend the vast majority of their lives underground eating worms, insects, and even small mice and amphibians. Tiger salamanders are the largest land salamander in North America, growing to about one foot in length. Their colors vary, but they generally have orange-yellow blotches on dark skin. The best way to see these creatures is by visiting a vernal pool in early spring right after a few days of rain.

As early as mid-January, tiger salamanders come out from underground tunnels and head to the vernal pools to breed. The salamanders gather in a wild courtship dance of twisting, nudging, and tail waving. Hundreds of salamanders squirm and wiggle together in the pool.

After the eggs are laid, the adults return to the forest. The eggs are laid in clusters about the size of a golf ball. In about one month, the eggs hatch, and a few months later the young reach maturity and leave the pool for their forest tunnels—until they return the next year.

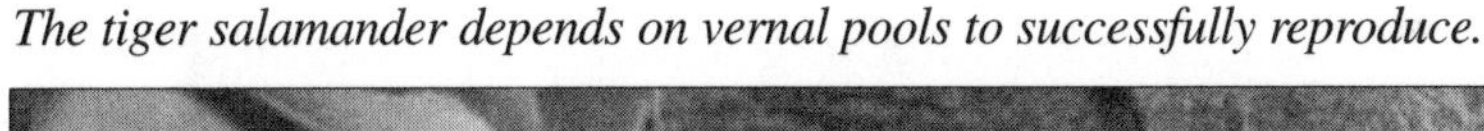

The tiger salamander depends on vernal pools to successfully reproduce.

by pollution, city officials had a choice. They could spend $6 to $8 billion on a water-filtration plant and spend $300 to $400 million in yearly maintenance, or they could use money to protect the wetlands and other lands around the reservoirs that filter the water naturally, at a cost of only $1.5 billion. Instead of using expensive technology, city officials chose to use the wetlands and other natural areas as the way to provide clean water for millions of people.

Groundwater recharge

Not only do wetlands clean and purify water, they can also be key places for collecting and storing drinking water. Drinking water comes from surface-water supplies such as lakes and reservoirs and from underground water supplies called aquifers. Aquifers are places underground where water is stored in spaces between the sand and gravel. Aquifers vary greatly in size and depth. To keep aquifers filled, water must soak into them through the ground in a process known as recharge. Recharge comes from precipitation that falls on the ground. Wetlands act as reservoirs, collecting water from the land that surrounds them. Wetlands then release the water they collect into above- and below-ground water sources.

Most of the recharge from wetlands takes place on the edges of wetlands. This makes small wetlands very effective in helping add water to aquifers and demonstrates the importance of even small wetlands. The water in wetlands is not only useful for humans; it is also what makes wetlands vital habitats for so many animals.

Animal life

Wetlands are also valuable because they are extremely productive and diverse ecosystems. They rival rain forests and coral reefs in the number and variety of animals and plants that depend on them. Wetlands are a haven for animals because they provide food, water, and shelter.

During a drought or dry season, wetlands may be an animal's only nearby food and water source. The plants in a wetland provide a source of food even after they die. The dead

Wetlands supply food, water, and shelter to animals as large as the moose.

leaves fall into the water and are decomposed by bacteria and other microorganisms. This mix of organic particles formed by dead plants and animals is called detritus. Detritus becomes food for small animals, which in turn become food for larger animals, and so on. This steady supply of detritus forms a base for an extensive food web.

Wetlands are home to a wide variety of animals, from moose, beavers, mink, ducks, and geese to smaller animals such as amphibians, fish, and countless invertebrates. Besides the animals that live in wetlands year-round, there

Whooping Cranes

One of the most spectacular birds in the world is the whooping crane. Standing five feet tall, it is the tallest bird in North America. Whooping cranes have a wingspan of eight feet and weigh up to sixteen pounds. They are snowy white, except for black wingtips and a reddish-black patch of feathers on their heads. Young whooping cranes are reddish-brown until they are about four months old, when white feathers begin to appear. Whooping cranes are the only large birds that fly with their neck and legs straight out.

These spectacular creatures are also one of the most endangered birds, because so many wetlands they depend on are now gone. In 1946, the whooping crane population dropped to just sixteen birds. Many wetlands were being drained or filled for agriculture throughout the Midwest, so the cranes had fewer places to breed. With the reduction in wetlands, there were also fewer places for them to rest and feed in on their migration south.

Before the whooping crane was lost forever, people began to help the whooping cranes recover by protecting key wetlands. This has been successful, but scientists still worry that the cranes could be wiped out by disease, drought, or hurricane. To prevent this, scientists are using captive cranes to establish a second group of cranes with a population of at least twenty-five breeding pairs. The scientists have even used small planes to teach the captive-born birds to fly the proper migration routes from one wetland to the next. Without the wetlands, the cranes and many other animals would not be able to survive.

White whooping cranes depend on North American wetlands for sustenance along their migration routes.

are also many animals that use wetlands on a temporary basis. Waterfowl and other birds use wetlands as resting spots on their migration routes. In the summer, many animals go to wetlands to breed because of the abundant food, water, and shelter there. About 80 percent of America's breeding bird population and more than 50 percent of the eight hundred protected migratory species rely on wetlands to carry out their life cycles.

Sandpipers are just one group of birds that are dependent on wetlands for breeding and migration. Jim Corven, director of the Western Hemisphere Shorebird Reserve Network, has identified three hundred important wetlands that are stopover areas from Alaska to Tierra del Fuego at the southern tip of South America. One such place is the Bay of Fundy in Nova Scotia. Don Stap writes in his article "Living on the Edge," "In some years as many as 1.4 million semipalmated sandpipers, possibly 95% of their world population, stop here [in the Bay of Fundy], where the world's greatest tides—they rise and fall as much as 50 feet—expose miles of food-rich mud flats."[6] The sandpipers can find the food they need in these wetlands to give them the energy for the thousands of miles they fly during their migration.

Endangered species

The crucial habitat that wetlands provide to many animals plays an important role in the survival of many endangered species. Animals become endangered for a variety of reasons. However, the most common cause is loss of habitat. By providing this habitat, wetlands help to maintain diversity, which contributes to a healthy environment. Forty-five percent of endangered species in the United States use wetlands for either all or part of their life cycles. A few of the endangered animals that make their home in wetlands are wood storks, Florida panthers, bog turtles, and manatees, large aquatic mammals that look like a cross between a seal and a whale.

Being home to so many endangered species has also benefited wetlands. When an animal is listed as endangered, the Endangered Species Act requires a plan for saving the animal, which often involves preserving the animal's habitat. Therefore, if endangered species live in a wetland, the wetland will be protected.

Food from wetlands

Wetlands provide food not only for animals but for humans as well. Humans eat a variety of food that comes from wetlands. According to the Environmental Protection Agency (EPA), the fish and shellfish that depend on wetlands for all or part of their life cycle make up more than 75 percent of the commercial and 90 percent of the recreational catch. (The U.S. fishing harvest is valued at $2 billion a year.) Oysters, clams, salmon, bluefish, and striped bass are just a few of the fish and shellfish that spend all or part of their lives in wetlands. The wetlands act as nurseries for most of these species. In the shallow waters and among wetland plants, young fish find abundant food, as well as protection from predators.

Nearly half of the endangered species in the United States, like the Florida panther, exist in wetland territories.

Fish are not the only food people harvest from wetlands. Blueberries are grown in wetlands in Maine, New Jersey, and other areas. Cranberries are grown in bogs in Massachusetts, New Jersey, and Wisconsin. Wild rice is found in the wetlands of the upper Midwest, and honey is made from many wetland flowers.

Some wetlands are used for food production in the same way that farmers use land to grow crops. This practice is called aquaculture, and when wetlands are used in this way, they are referred to as managed wetlands. Rice is grown in managed wetlands in Asia and is a major source of food for people all over the world. In the southern United States, crayfish are raised in the same wetlands with the rice. When managed for food production, wetlands do not function in the same way they would if left alone. As production increases and the wetlands become managed by humans to a greater extent, their natural functions decrease.

Other resources

Another resource that comes from wetlands is timber. Wetland forests provide millions of acres of timber for firewood, newsprint, and lumber. Most wetland timber in the United States is found east of the Rockies, primarily in Mississippi, Louisiana, Florida, and Georgia. Atlantic white cedar and loblolly pine are the trees harvested the most. Atlantic white cedar is used for boat building and docks. Another tree harvested for lumber that grows in wetlands is the red maple, which is commonly used for firewood. Of the 42 million cubic feet of cypress harvested per year from wetlands in Florida, about half is used for lumber; the other half is used as mulch for landscaping.

Peat moss is another wetlands resource that people use. Most of the peat comes from Russia and Canada. It is gathered in bogs and then dried and sold for use in home gardens. Gardeners use peat moss to retain water in the soil. Peat is also burned as fuel to heat homes in some countries of northern Europe. This use has declined over recent years in favor of other fuels, but increased amounts of peat are sold for horticultural uses.

Wetlands can offer unique, easily accessible opportunities to bird-watch, take photographs, or hike along trails.

Recreation

Wetlands are also popular places for tourism, recreation, and education. According to the EPA, half of the adults in the United States hunt, fish, bird-watch, or spend time photographing nature. Wetlands are an ideal setting for all these activities. About 98 million people spend approximately $59.5 billion per year on outdoor activities.

Since wetlands are home to such a wide variety of waterfowl, they are popular places for hunters. Up until the early 1900s, hunting was a major threat to wetlands. The birds there were hunted for food, and their feathers were used for hats. Today, state regulations limit the number of waterfowl that can be hunted, which helps to maintain duck populations without denying hunters their pastime.

Many people like to visit parks and nature preserves in wetlands. In the Everglades National Park alone, 600,000 people visit the Anhinga Trail, an educational nature trail through a wetland, every year. Wetlands give children and adults a chance to enjoy interesting wildlife and plants and at the same time educate them about ecology. As tourist attractions, wetlands can also help local economies by attracting visitors, who in turn spend money at local businesses. Hunters in search of waterfowl, for example, spend $600 million on supplies, lodging, and food annually.

Research

Wetlands serve as important sites for research in many areas. One area of study that particularly benefits from wetlands is archaeology, the study of prehistoric life and cultures. Because of the lack of oxygen in wetland soil, the bacteria that normally decompose organic material cannot survive. So, there is little decay in some places. In these areas, human and animal remains are sometimes found in an amazingly well-preserved state. Archaeologists can learn a lot about life in the past from these remains and other artifacts they find. Scientists studying climate change can also learn about the past by examining ancient pollen spores found deep in bogs. Using the pollen, scientists can identify the plants that were growing in the area in ancient times and determine the climate changes that have taken place throughout history.

Wetland plants are often involved in scientific research. Over the years, these plants have provided various medicines for human use. For example, willows are the original source of salicylic acid, which is the active ingredient in aspirin.

Global benefits

Wetlands also play a role in biospheric stability, preserving the thin skin of air, water, and soil that surrounds the earth where all life lives. Wetlands contribute globally by maintaining air quality.

Wetlands emit and/or absorb the chemicals that make up the air we breathe. Air is a mixture of gases, and nitrogen and oxygen make up 98 percent of them. Some of the other gases in the air include methane, water vapor, and carbon dioxide. Wetlands help maintain the proper balance between all these chemicals by hosting large communities of plants that play a role in the oxygen cycle. This helps maintain the right amount of oxygen in the atmosphere.

Bogs and other wetlands primarily made from peat play an important role in maintaining carbon levels in the atmosphere. When excess carbon gets into the atmosphere from the burning of fossil fuels and deforestation, average temperatures rise around the world. This phenomenon is called global warming. The plants in bogs can help counteract global warming, however, by taking in carbon dioxide during photosynthesis. Since there is very little decomposition in the bogs, the carbon builds up and remains in the bog as peat. This carbon transfer from the atmosphere to the bog helps to balance the increase in carbon caused by human activity. Bogs, therefore, act as a sink to absorb excess carbon.

According to Steve Rice, a botany professor at Union College,

> One important issue for the future is that if global warming raises temperatures high enough, there will be more decomposition in the bogs. This will cause more carbon to be released into the atmosphere. With more carbon released there will be higher temperatures, which will lead to more decomposing and more carbon being released. This interaction may create a cycle that could permanently alter the world's climate.[7]

Measuring the value of wetlands

Wetlands are invaluable to the earth's health because of their ability to maintain the earth's climate and their interrelationships with other ecosystems. Wetlands are also living laboratories, and scientists are just realizing the critical role they play in the environment. This is why the threats to wetlands are so serious; they pose problems not only for the wetlands themselves but for all life dependent on wetlands, including humans.

3

Wetland Destruction

THE EXACT RATE of wetlands loss around the world is unknown. Accurate records are not kept in many places, and much of the loss of wetlands occurred centuries ago. It is believed that roughly half of the original wetlands in the world have been lost. Some places have lost more. For example, in New Zealand, 90 percent of the wetlands are gone. Britain has lost 60 percent of its wetlands. These losses have caused a variety of problems, from increased flooding and water pollution to the loss of habitat for countless animals and plants.

The Emergency Wetland Resources Act of 1986 requires the Fish and Wildlife Service to study the status and trends in the wetlands across the United States. Each decade, it must report the findings to Congress. According to the latest report in 1997, there were an estimated 105.5 million acres of wetlands in the lower forty-eight states. This is about half of what was there in the 1600s. In the United States, most of the loss occurred from the mid-1950s to the mid-1970s, an active time for industrial development. The total area of lost wetlands is about the size of California. Only Alaska has about the same amount of wetlands now as it did then, approximately 170 million acres. Fortunately, the rate of wetland loss is now decreasing in the United States. The report found that the loss of wetlands is now about 58,500 acres per year, which is an 80 percent drop from a decade ago.

The diverse threats to wetlands, however, are still widespread. Many places where people work, live, or play were

Wetland Loss

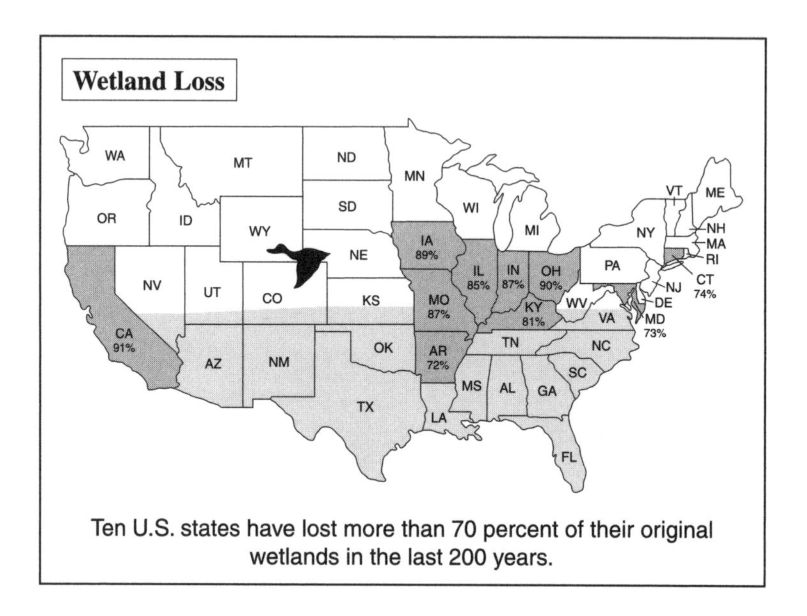

Ten U.S. states have lost more than 70 percent of their original wetlands in the last 200 years.

once wetlands that have now been completely changed. Looking at modern houses, offices, airports, and parking lots, it is hard to imagine the wetlands that existed there before.

Some of the problems wetlands face are caused when humans destroy the wetland and replace it with farms, buildings, roads, or some other development. Wetlands are desirable farmland because the land is flat, has good soil, and has available water. Wetlands are also attractive for development because there is usually nothing else built on the land. Coastal and lakeshore wetlands also have the advantage of providing scenic views for the homes that are built nearby.

Consequences of wetland loss

As wetlands are degraded in preparation for development, the habitat they provide becomes fragmented into smaller pieces. The habitat fragmentation causes many problems for animals and plants. The loss of habitat makes it harder for large wide-ranging species such as moose to find enough food, water, and shelter. Having less wetlands area means that there is more competition for fewer resources. Another consequence is that, as development encroaches on wetlands, it becomes easier for diseases and parasites from domestic animals and plants to come in contact with the remaining wild plants and animals. All these problems and more are caused simply by fragmenting a wetland into smaller areas.

Smaller-scale changes can also affect the quality of life in a wetland. For example, as more homes are built in forested areas, they have subtle effects on the local environment. Driveways and roads with high curbs become huge barriers to salamanders that live in the forest but return each spring to vernal wetlands. Since they cannot reach their breeding grounds, their population declines. This loss of one species threatens the overall health of the wetland, which is dependent on interrelationships among its inhabitants.

Boston's Wetlands

When Reverend William Blaxton set up his farm on the Shawmut Peninsula in what would become Boston in 1625, he had no intention of starting a colony, much less one of the largest cities in the United States. But when the Puritans came in 1630, things changed quickly. The Puritans believed that they were destined to build a perfect community based on their belief in God. Part of their beliefs was that the land was given to them by God to create their community. They began right away.

A marsh nearby provided building materials for their homes and plentiful food for both their livestock and their tables. By 1640, the population had risen to twelve hundred. In 1641, the first marshes were dug up and filled in. By 1645, all the marshes had either been filled in or changed in some way to serve the growing population. Some were turned into stagnant ponds that caused a terrible stink through town, which led to the filling in of even more wetlands. The famous Fenway Park, home of the Boston Red Sox baseball team, sits on what used to be a wetland. In fact, the word *fen* is the name of a type of wetland.

Humans also feel the effects of wetland loss. Before European settlement, the swamp forests along the Mississippi could hold sixty days' worth of floodwaters. Now the storage capacity is just twelve days. This had serious consequences in 1993 when, after two months of heavy rains, there was extremely heavy flooding in the Upper Mississippi Valley. The flood left tens of thousands of people homeless, caused $20 billion in damage, and killed at least thirty-eight people. It was the most expensive flood in U.S. history. Over the years, 20 million acres of wetlands in the area had been drained and were no longer there. If 13 million acres had remained, it would have been enough to prevent the devastation.

The threats facing wetlands do not work in isolation. Because of the intricate interrelationships that make up any ecosystem, a problem in one place has an effect on other parts of the system. For example, when a forest is cut down for lumber, there is usually increased erosion because there are no tree roots to hold the soil. The soil washes into nearby streams and can be carried into wetlands. Too much soil in a wetland will begin to fill the wetland in. In addition, a wetland is usually under pressure from more than one threat; for example, a new housing development may fragment the wetland into smaller pieces and add pollution to the water at the same time.

Because of water's importance in wetlands, changes in the water level severely affect them. Increases in water levels can damage and destroy wetlands by turning them into ponds or lakes or, in the case of coastal wetlands, forcing them to become part of the ocean. Decreases in water will dry the wetland out. The amount of water changes the soil, which then affects the plants, which in turn influences the animals. In the end, the entire wetland is altered.

Filling wetlands

Wetlands are changed when they are converted to dry land by being filled in or dumped on. Wetlands have historically been considered wastelands because of their appearance, and garbage, dredged materials, soil, and other waste have been dumped on wetlands. This type of dumping buries wetland soil and lowers the water table.

Material is also dumped on wetlands to develop and completely alter the site. Tons of rock and soil may be dumped on a wetland to fill it in and change the way the land is used, which permanently destroys it. Large parts of many major coastal cities such as New York, Boston, Philadelphia, and Washington, D.C., were built on wetlands. Inland wetlands are also being filled in and developed. Wetlands are also now prime areas for building new roads, airports, and industrial sites.

Besides the outright destruction of wetlands, development can have negative effects on any adjacent wetland in the surrounding area. The construction of roads, buildings, parking lots, and other development means that when rain or snow falls, the water does not soak into the ground and slowly move into a nearby wetland. Instead, the water washes over the surface of concrete and rushes into the wetland faster than it can be absorbed. The water often carries pollutants and excess sediment. In addition, the water is warmer than usual. Warm water has less oxygen, which means there is less oxygen available for animals.

Draining wetlands

Wetlands are also commonly drained as a way of readying them for agriculture, development, or mosquito control. Millions of acres of wetlands have been drained for these purposes, leaving them completely dried out. It is hard to believe, but Central Park in New York City was once a wetland. In 1858, Colonel George Waring drained the area, declaring it a menace to public health. Many people in those days believed that wetlands were full of diseases and dangerous animals.

Draining wetlands has occurred since people first established living areas near wetlands. They were originally drained by hand-dug canals. Draining wetlands increased in the late 1800s when better soil-drainage technologies emerged. The use of tile drains, pipes, and mechanical ditch-digging equipment allowed faster and

Draining wetlands for agriculture and development was once an acceptable practice, but it has recently been shown to be environmentally catastrophic.

easier draining. Places that were once too wet for farming could be converted to usable farmland. States were encouraged to drain wetlands to make more farmland available. These former wetlands became very productive farmland. In fact, some of the best farmland in the world comes from wetlands that were drained of their water in Illinois, Iowa, and southern Minnesota.

Wetland development was especially heavy after World War II as the population grew and development boomed. It became fashionable for people to move into single-family homes, which required new places for construction. The advent of air conditioning and the use of pesticides helped to make building a home in a former wetland a lot more feasible; the heat and animal life of wetland areas would no longer be obstacles to people who wanted to live there. The application of pesticides, however, killed not only mosquitoes but millions of other animals in the wetlands. With more houses came more roads, which also damaged the wetlands. Coastal wetlands were particularly hard hit by urban and industrial development.

This development continues today. As more houses are being built in forested areas, vernal wetlands are being filled in. The destruction of small wetlands like vernal pools can be especially disastrous. According to John Turner, director of conservation programs for the Long Island chapter of the Nature Conservancy, "There is increasing evidence that isolated small seasonal wetlands are critical to [the] survival of a host of wildlife species,

Draining Innovations

Draining wetlands was hard work, much of it done by hand or simple machines. Ditches were dug so that water would flow out of the wetland, but the water would often soak back into the ground at the bottom of the ditch. It was not until 1835 that people began placing ceramic tiles at the bottom of the ditches to keep the water from soaking back into the ground. Tiles were first imported to the United States by John Johnston, a farmer in Geneva, New York. Johnston had a farm with poor drainage. Water just sat in what could be fields. After reading about tiles used in Scotland, he bought some for his farm. The tiles were used to underlay ditches and help water drain out faster and more efficiently. Once neighboring farmers saw how well the tiles worked, they also wanted tiles for draining their land. Johnston and his friend Benjamin Whartenby built a factory to make inexpensive tiles. As the use of tile draining spread, so did tile manufacturing.

Within a generation, the farms had changed so much that places that were once wetlands no longer supported wetland plants. They became more fields for farming. Even the ditches were no longer needed; the water was gone. After one generation, it was easy to forget where the tiles were buried without consulting written plans.

Tiles were still being laid by hand or with horses until 1892, when steam-powered ditch-digging machines were invented. These ditch diggers could cut a ditch up to 1,650 feet long and 4.5 feet deep every day, making it easier to lay tiles. Steam-powered diggers were followed by the gasoline-powered ditch digger in 1908, which was even more powerful. Huge areas of wetlands could be drained of their water. Between 1906 and 1922, nearly 9 million acres were drained in seven states. In just sixteen years, Illinois, Indiana, and Iowa lost 30 percent of their wetlands. Land previously too wet to farm was now available for farming. Although these farms produced food for a growing nation, wetlands were disappearing.

mainly amphibians, such as frogs, toads, and salamanders. These wetlands provide unique environments where conditions are just right for amphibians because there are no predators like fish, which live in larger wetlands."[8] These small wetlands and the wildlife they support are in danger.

Draining wetlands is often done as a method of mosquito control. Wetlands are home to mosquitoes, which can carry diseases such as malaria and the West Nile virus. Many acres of wetlands have been victimized by efforts to curb mosquito populations. Mosquitoes need stagnant water for successful reproduction because their larvae live just under the surface of the water. The slow-moving water of a wetland provides the perfect habitat for breeding mosquitoes. Draining the water effectively destroys the mosquitoes' breeding grounds, but in the process it also damages the wetlands. Other techniques to eradicate mosquitoes have included building canals that bring in salt water, which kills the larvae but also kills other things in the wetland. Since

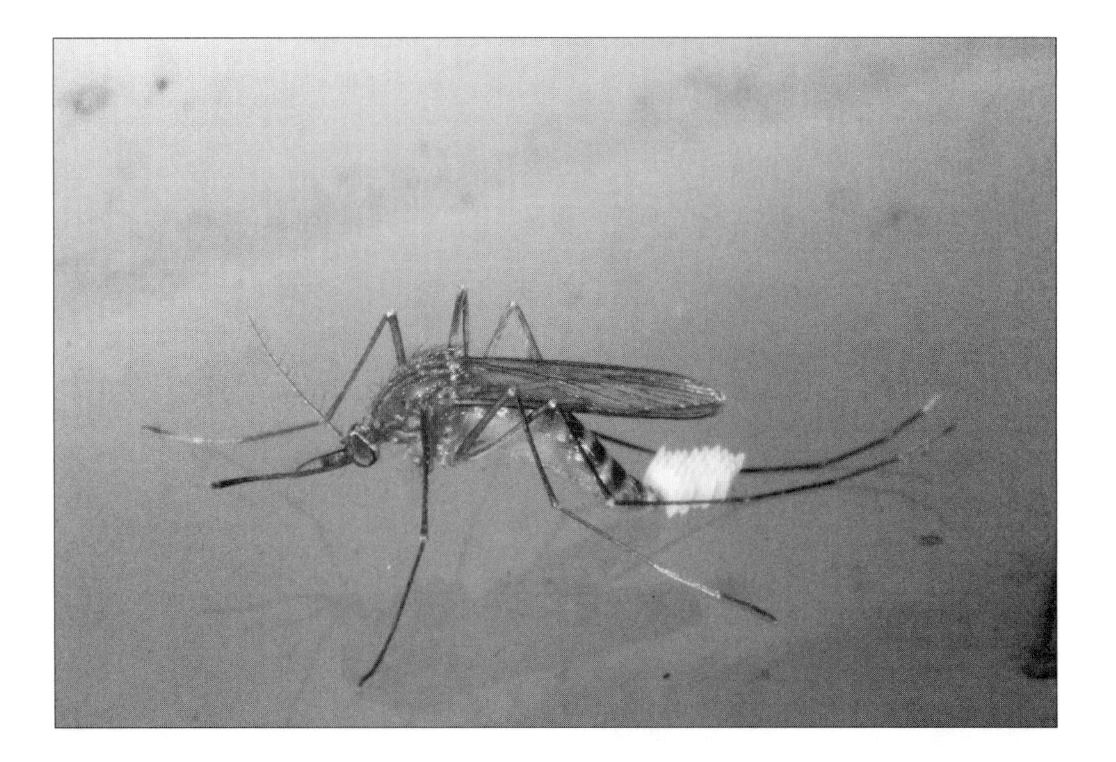

Stagnant water found in wetlands presents an optimum habitat for breeding mosquitoes.

mosquito larvae live just under the surface of the water, oil is sometimes spread across the surface to cut off the oxygen supply to the larvae.

Although these practices did help reduce the spread of malaria and other diseases, many acres of wetlands were destroyed in the process. Beginning in the 1970s, the harmful effects of draining wetlands to remove mosquitoes became clear. As a result, less damaging techniques are now used to control mosquitoes. One method uses minnows and other small fish to eat the mosquito larvae. Another is to build nesting boxes that attract mosquito-eating animals such as bats and swallows.

Water use

The large amounts of water people use for farming, manufacturing, drinking, and recreation threaten many wetlands. Water is pumped out of the ground for industrial and domestic use. Even snow-making operations at ski slopes pump out water from nearby streams. When too much water is taken from underground sources, the water table of the surrounding area can be lowered. With a lower water table, wetlands are prone to drying out. This threat is not obvious at first, and it takes a while before the problems are evident. It may not even seem like the wetland has been degraded unless one looks carefully at the changes in plant life caused by the drier conditions. Slowly, the place becomes a different habitat, and the wetland values are lost.

Flood-control measures may also isolate wetlands from their water sources. For example, levees built along rivers to prevent overbank flooding end up isolating wetlands from the river water they need. The wetlands then become drier and drier until they eventually become a field.

Dredging wetlands

Wetlands also face problems when their soil is taken away. Machines dig up the soil under wetlands to create channels of deeper water. The removal of materials from wetlands is called dredging. There are many reasons wetlands are dredged. Dredging is done to create channels deep enough for boats to navigate. This is done for both recreational and commercial boating. One large area of dredging has been the Intracoastal Waterway, which extends about three thousand miles through salt marshes, bays, and rivers on the Atlantic coast from New York City to Brownsville, Texas. Dredging is also done for industrial use. In the swamps around the Lake Maurepas and Lake Pontchartrain area of Louisiana, large canals were built for oil exploration.

Many wetlands are crisscrossed by man-made canals that were dredged for flood control. The goal of these canals is to drain the land surrounding the wetlands faster, but the canals damage the wetlands in many ways. Normally, water moves slowly across the entire wetland like a wide, shallow river. By speeding the flow of water through the wetland and funneling it to certain parts, areas of the wetland become dried out. In addition, when canals are built for boat traffic, the wave action from the boats causes erosion and damages plant life. Increased erosion causes increased sediment in the water, which lowers the clarity of the water, making it harder for plants to obtain enough sunlight. In coastal areas, dredging and canals can cause saltwater intrusion, which changes the type of wetland from freshwater to salt water, which then changes the plants and animals that live there.

Flooding wetlands

Although removing water through draining and dredging is a critical problem for wetlands, too much water can also threaten them. When water levels in a wetland rise, the area is no longer a wetland; it becomes a pond or part of a lake. The presence of dams increases this threat. Dams are put in place to block streams and rivers in order to generate electricity and control recreation, irrigation, and flooding. The dams create ponds, lakes, or reservoirs. When these bodies of water are created, they flood wetland areas upstream from the dam and destroy them.

One example of how dams or water-impoundment projects damage wetlands can be seen in northern New Jersey

Muskrats typically build their burrows in cattail marshes.

and the Pocono Mountains of Pennsylvania. Over time, wetland streams were blocked there to create lakes for recreation and water supplies. Doing this created a new habitat, and many of the wetland plants and animals were not able to survive the transition. For example, muskrats, which are small beaverlike rodents, prefer cattail marshes where they can find plenty of food and plant materials to build their dome-shaped homes. When a wetland becomes a pond or lake, the deeper water will not support as many cattails, which in turn limits the muskrat population.

Coastal wetlands are normally protected from a slight rise in sea levels because they contain a buildup of sediment and an accumulation of dead plants. The presence of these items is dependent on the right amount of flooding from rivers and the ocean. Flooding from rivers carries in sediment, and high tides from the ocean bring in materials that help to build up wetlands.

This balance can be disrupted by human activity. Canals, dams, and other human changes to rivers reduce the sediment flow that would normally raise wetlands to balance the changes in sea levels. When flood-control measures prevent waters from overflowing, less sediment is laid across the wetlands along the coast and the land sinks. In Texas, for example, the underground mining of oil and gas has caused the land surface to drop. The withdrawal of water from aquifers also causes subsidence. In Galveston Bay, Texas, the level of the marsh dropped two and a half meters because of groundwater and mineral extraction. Dredging in many wetlands, especially around large coastal cities such as Houston, causes subsidence.

Subsidence is a serious problem. In Louisiana, $300 million of coastal real estate could be affected over the next fifty years. The saltwater intrusion that results from subsidence could impact fisheries and trapping in Louisiana and threaten a billion-dollar industry. With the threat of higher sea levels from global climate change, subsidence will become an even greater problem.

Climate change

One threat to wetlands that is becoming increasingly serious is the rise in temperatures around the world, known as global warming. The use of fossil fuels and deforestation have increased greenhouse gases in the atmosphere. These gases hold in increased amounts of heat from the sun, which raises temperatures. Higher temperatures cause two problems that lead to a rising sea level. When heated, all matter expands. Higher temperatures will expand seawater, which means the water will need more space and will creep farther up the shore. Warmer temperatures will also melt ice caps at the North and South Poles, leading to more water in the oceans. As a result, the sea level will rise and drown coastal wetlands. Global sea levels have already risen ten to twenty-five centimeters over the past one hundred years. And sea levels are expected to continue to rise in the future. Although there is some debate on the impact of global warming, it is becoming more and more clear that global warming is a reality.

The Force of Fires

Fires are a force that can both benefit and damage wetlands. For many wetlands, fires have always been a natural part of the ecosystem. The water in a wetland usually keeps fires from becoming too severe. Many wetland plants depend on fires to aid in their reproduction. Pitch pines need the heat from fire to release the seeds from their cones. Fires in cedar and cypress swamps clear out brush and shrubs and give seedlings a better chance to grow. Some plants rely on fires to create burn holes; these holes allow certain wetland plants the chance to grow without competition for sunlight and space.

However, as people drained wetlands of their water and dried them out, fires became more frequent and more severe. The Everglades have been particularly hard hit by the increase in fires. These fires do more than burn plants; they burn the organic matter in the soil. The fires can eventually burn off enough soil to expose the limestone bedrock that lies underneath the Everglades. When the soil is lost, changes occur to the ecosystem. New plants move in, taking advantage of the changing conditions. The fires also flatten the landscape by burning off the slight changes in elevation that once created a more varied habitat. Because of human actions, fires have gone from being a force that created diversity in wetlands to a force that limits the variety of life.

Wetland fires can produce positive organic results when burning is moderate, but more severe blazes have disastrous effects.

In a report to the United Nations in June 2002, the Bush administration, for the first time, blamed human actions for global warming and its consequences. According to the report, it is "very likely"[9] that the United States will see the permanent disappearance of coastal marshes due to rising waters from global warming.

Despite the debate on the specific effects of global warming, higher temperatures will most certainly cause problems for freshwater wetlands too. Global warming will affect the amount of precipitation, which provides the water supply for inland wetlands. Some wetlands will dry out and others may expand.

This loss of wetlands will have serious ecological and economic effects. Plants and animals of coastal wetlands will lose the habitat they need for survival. There will be increased damage to the wetlands from storms and hurricanes. The fish and shellfish industry will be financially impacted. Homes and buildings near the coast will endure more damage from erosion and storms. Most important, the functions of these wetlands will be lost.

4

Wetland Degradation

SOMETIMES IT CAN be difficult to realize a wetland is in danger because the forces that threaten to change it do not completely destroy it. Although the wetland may still exist and even retain much of its beauty and functionality, a close examination will reveal its problems. These damaging forces can include pollution, lumbering, mining, and threats by non-native plants and animals, which all serve to degrade the wetland to the point where its survival is threatened.

Pollutants

Wetlands can be thought of as the kidneys of the earth, filtering water and releasing it back into streams, rivers, and other bodies of water above and below ground. Unfortunately, wetlands become more like failing septic systems when too many pollutants end up in wetlands and their system cannot handle them.

Pollution comes from a variety of sources, such as industrial waste, landfills, oil spills, road salts, runoff from farms, sewage treatment plants. Pollutants can enter wetlands from hundreds of miles away and in two different ways: from point and nonpoint sources. A point source is pollution that comes out of a pipe or some other direct means into the wetland. A nonpoint source is pollution that washes into the water off things such as roads, lawns, and golf courses. Both types of pollution can be a problem, but it is much harder to control nonpoint pollution because it can wash in from anywhere.

A biologist examines the damage done to Devil's Swamp, near Baton Rouge, Louisiana, from recent point source pollution.

Many landfills and factories have been built on or near wetlands, exposing them to heavy metals and toxic chemicals from industrial waste. Pollution from the landfills can leak into the water and end up in the wetlands. Other pollutants such as oil can end up in a wetland from both large and small spills. Pesticides and herbicides find their way to wetlands when people apply them to lawns, farms, and golf courses to kill insects and unwanted plants. Rain and snow then wash the pesticides and herbicides into the soil, where they eventually work themselves into the groundwater that leads to wetlands and other bodies of water.

Pollution that threatens wetlands can even come from the sky as acid deposition, also known as acid rain. A variety of chemicals from automobile exhaust, electrical generation, and other industrial activities float up into the sky. There, these chemicals combine with water molecules. When these polluted water droplets fall into wetlands as rain, snow, or sleet, the water becomes more acidic. Many of the pollutants that end up in acid rain are generated in the midwestern United States, and winds carry them eastward, where they eventually fall on wetlands, lakes, and forests in the Northeast.

Although wetlands can tolerate and filter out a certain amount of pollutants, if enough pollutants enter a wetland, they can kill off plants and animals very quickly. Depending on the type and quantity of pollutants, animals may even die from immediate contact. For other pollutants, the effect takes time but is no less devastating. Pollutants cause deformities, reproductive problems, cancers, and death in aquatic animals. And because of the extensive food chains found in wetlands, pollutants spread very quickly. For example, when an aquatic insect eats a plant that has a pesticide on it, the chemicals become part of the insect. When a small fish comes along and eats the insect, it is also ingesting the pesticide. Each time a bigger fish comes along and eats these smaller fish, it ingests more and more pesticides. The concentration of a pesticide magnifies as it moves through the food chain.

The pesticide DDT (dichlorodiphenyltrichloroethane) has a history of polluting wetlands. In the 1950s and 1970s, DDT was used to kill mosquitoes and many other insects. It entered wetlands when rain washed it off the surrounding lands or it was sprayed directly on wetlands to kill mosquitoes. The disastrous effects of DDT went all the way through the food chain. For example, when ospreys (a fish-eating hawk) ate the fish that had swum in waters polluted with DDT, they accumulated dangerous amounts of the toxin in their bodies. This damaged their ability to lay eggs, and the osprey population plummeted. However, with the ban on DDT in 1972 and other efforts to reestablish nest sites, the osprey have come back to many wetlands.

Nutrients

Excess nutrients like nitrogen and phosphorus can also threaten a wetland's health. While wetlands can handle and clean up certain levels of nutrient pollution, they do not have an unlimited capacity. Too many nutrients lead to an increase in plant growth. This process is called eutrophication.

Osprey populations that plummeted in the 1950s, when pesticide pollutants appeared in wetland food chains, have begun to recover.

When the plants die, there are too many dead ones for decay to take place fast enough to keep the wetland from filling in. So, the wetland fills with soil, dries out, and becomes a field or meadow.

The coastal wetlands and parts of the Gulf of Mexico are threatened by excess nutrients, which flow down the Mississippi from farms all across the Midwest. Many things can be done to prevent this excess of nutrients, including changing crops on some farms, using less fertilizer, and increasing water treatment. However, the most effective solution—and the least disruptive to farmers—is to restore wetlands along the river so that the excess nutrients can be filtered out before they reach the Gulf of Mexico. These restored wetlands would also help ease flooding problems along the Mississippi River.

Mining

Mining is another process that harms wetlands. When areas are mined, materials are extracted from them. Although the mined resources have many uses, the work to obtain them can damage wetlands. For example, peat is mined from bogs for fuel in electrical generation, as well as for horticultural and agricultural uses. Most of the peat comes from Russia and Canada, the United States, New Zealand, and northern Europe. Peat mining can destroy a bog and adjacent wetland areas because it requires the bog to be drained and then machines move in to dig out the peat. In a

Railroads and Wetlands

New inventions can have unintended consequences on the environment. Railroads, for example, have had a widespread effect on wetlands. Railroads changed the way food and other natural resources moved around the United States. Before railroads, farmers were limited to markets close to where they grew their crops. With railroads, food and other materials could be transported over great distances, even across the country. This was met with a growing demand in urban areas for these resources, which made farming more profitable. Land that was too wet to farm was now worth the expense of draining.

For example, new railroad routes and the opening of the Panama Canal meant that food grown in California could reach the big cities on the coasts. In California, 700,000 acres of wetlands became farmland by 1918. By 1922, 70 percent of the state's wetlands were transformed to meet the growing demand for crops that railroads helped to create.

Forested swamps also suffered from the demand for resources spurred by railroads. Lumber from forested swamps could be transported long distances and sold to build houses for a growing population. Railroads themselves used huge amounts of wood for ties, bridges, and fuel. In Ohio, the railroads used 1 million cords of wood a year. To meet the need for all this timber, farmers cut down enormous numbers of trees in the Black Swamp. The Black Swamp once covered an area 160 kilometers long and 40 kilometers wide in northwestern Ohio. Except for a few small areas, this huge swamp is now gone.

similar way, sand and gravel mining along the coastal plain areas can degrade salt marshes. Phosphate mining from wetlands in central Florida has also caused the loss of thousands of acres of wetlands.

Although coal is not mined from wetlands, coal mines are sometimes located near enough to wetlands that they are affected by the mines' pollution and waste. The tailings (leftover material from mining operations) are often dumped near wetlands. Acid drainage from these metal-rich tailings causes extensive damage to wetlands by making the water too acidic for fish and other animals to survive. The negative effects of mining have decreased with better wetland protection, but the threat remains present and conflict over mining practices continues.

On May 3, 2002, the Bush administration approved a regulation that will allow mining companies to dump leftover dirt and rock from mountaintop mining into streams and valleys. This method of dumping is less expensive and more convenient for the mining companies but devastating for wetlands. The regulation has sparked a conflict between the mining industry, which claims that the regulation will ultimately save jobs, and environmentalists concerned with wetland issues. Though some legislators and experts think this allowance will not be problematic for the wetlands, others disagree. Joe Lovett, a lawyer in Lewisburg, West Virginia, who has brought lawsuits to end mountaintop mining, called the new regulations "a death warrant" for the environment and local towns. According to Lovett, "[The decision] opens up the coal region to more exploitation by the industry."[10] The *New York Times* points out in an editorial that "Dumping wastes in hollows and streams is certainly the cheapest means of disposing waste. But that doesn't make it right."[11]

All mining, however, does not destroy wetlands. Sometimes mining can actually create wetlands. There are some abandoned mines that harbor unique plant and animal communities. Nevertheless, when mining wastes are dumped in wetlands, they pollute them and change their natural state.

Lumber

Lumbering, the process of cutting trees, is also common in wetlands and can cause many problems. Lumbering is a major operation that requires a number of steps. Water must be drained away so that roads can be built in order to haul logs out. Damage is also done when the trees are dragged across the ground. When the wetland is cleared of trees, habitat is lost to the animals and plants that lived there.

The damage resulting from lumber operations may be temporary or permanent depending on the type of lumbering. With care, wetlands can be harvested for lumber with limited damage to the environment. Wetlands can recover

Pull Boats and Cypress Trees

Cypress trees have been harvested on a small scale ever since the first settlers came to the southeastern United States. The trees were initially cut with hand tools and floated out of the swamps during the spring floods. This meant that only trees growing near rivers could be harvested. The lack of technology protected most of the cypress trees.

This all changed after the Civil War when demand for timber grew. As a result, the value of cypress increased. Soon after, technology to harvest more trees was developed. In 1889, William Baptist of New Orleans invented a pull boat. These boats had steam engines attached to cables that could pull cut logs to the river. Logs as far as five thousand feet away from a river could be pulled by the boat and floated away. To keep the logs from getting tangled while they were being pulled through the forest, the tip of the log was fitted with a metal cone. The forest floor in many places was destroyed because of the huge logs that were dragged across it.

In just thirty years, most of the old-growth cypress in the southeastern United States was gone. For example, more than 1.5 million acres of forested swamps in Louisiana were cut down, along with the homes of countless animals, including the now extinct ivory-billed woodpecker.

from selective cutting when only some trees are taken and efforts are made to replant trees. On the other hand, wetlands suffer greatly when all the trees are cut at once in what is known as a clear-cut.

Grazing

Yet another activity that severely alters wetlands is cattle grazing. Because the most inexpensive land that cattle owners can lease from the federal government for their herds is usually quite dry, the cattle go in search of streams and wetlands for the water and grass that they provide. This concentration of cattle in confined areas begins a cycle that harms the wetland. To begin with, the cows' urine and dung pollute the water and earth. In addition to the heavy consumption of grass that the cows eat, animal dens and tunnels below ground often collapse as the huge animals walk, compacting the earth. According to Mike Matheson, president of the Native Utah Cutthroat Association, which works to restore streams and wetlands in Utah,

The waste products of cattle contribute to many problems in wetland regions.

The Death of a Wetland

In California's Central Valley, 95 percent of the wetlands have been converted to farmland. The wetlands that remain are crucial to birds migrating through California. One of these wetlands was preserved as the Kesterson National Wildlife Refuge.

In 1983, biologists at the refuge began to notice many dead and deformed birds. They discovered that there was a chemical called selenium in the runoff from the farms surrounding the refuge. The runoff had begun to contaminate the marshes in Kesterson. The refuge became a death trap for migrating birds. They saw a wetland they thought they could rest in but instead became poisoned. As a result, the Kesterson refuge was closed in the mid-1980s. The wetland was so polluted that it had to be drained to ensure that wildlife would stop coming there. It cost tens of millions of dollars to close and clean the site.

"The cattle compact the earth to such a degree that it creates a zone where no plants can grow. Additionally, the cattle add sediment to the wetland, damaging aquatic life."[12] Cattle grazing reduces streamside vegetation, which normally filters the water; unfiltered water contains more pollutants. Also, the reduction in plants means that there is less shade available, which leads to a higher water temperature. Water with higher temperatures can hold less oxygen, making life difficult for many fish. Efforts are being made to restore wetlands by fencing them off from cattle.

Invasive plants

Wetlands face serious threats from plants and animals that do not belong in them. These plants and animals are called invasive species, and they grow out of control and damage wetlands. Most invasive plants have been introduced to wetlands accidentally or through carelessness. Some are planted as garden plants and spread from there. Others are brought in for agriculture and other uses and spread to the wild. Many make their way to new places by accident, unintentionally spread by people. As humans travel more and more, the spread of invasive plants becomes easier and easier.

Invasive plants cause problems because they grow very rapidly and compete with native plants for sunlight, water, and soil. They are usually very adaptable and aggressive. Many invasive plants are normally adapted to live in poor soil and can survive extreme conditions. Since they are new to the wetland ecosystem, there are few, if any, herbivores (plant-eating animals) to keep the invasive plant populations in check. With no limit to their growth, invasive plants can survive easily and take over an environment quickly.

Invasive plants can cause problems for native wildlife. The plants take over and force out other plants, which limits the variety of plant life. This means there are fewer food options for animals in the wetlands. Because the wetland wildlife has not evolved with these new plants, there are few animals that can take advantage of the invasive plants for food. All of this reduces the number of species of animals the wetland can support. The Nature Conservancy estimates that 42 percent of the endangered and threatened species found in wetlands are in peril because non-native plants are upsetting the delicate balance maintained naturally. As David Pimentel, an ecologist at Cornell University, says, "It doesn't take many troublemakers to cause tremendous damage."[13]

One example of an invasive plant that is causing problems in wetlands is purple loosestrife. Purple loosestrife grows about three or four feet high with a beautiful spike of small purple flowers. It can cover wide areas of wetlands in a stunning display of purple. This plant was introduced by beekeepers and gardeners in the 1880s as a source of nectar for honey and for its use in gardens. Now, however, it is invading freshwater marshes, causing damage in forty-eight states. Purple loosestrife blooms in mid-July and August, producing millions of seeds per plant. Forty-five million dollars a year is spent to keep it from spreading.

Purple loosestrife is responsible for pushing out forty-four native plants and the animals that depend on them.

Invasive plants, like purple loosestrife, can irrevocably disrupt the delicate balance of wetland life.

According to an article by Dan Carrol in the August 2000 issue of the *New York State Conservationist*, "The abundance of purple loosestrife in New York's non-wooded wetlands might explain why marsh-dependent birds such as the black tern, least bittern, American bittern, pie-billed grebe, Virginia rail and sore have declined in the last 30 years."[14] To add to the problem, wetlands wildlife rarely eat the purple loosestrife seeds or leaves. With no natural predator, they grow dangerously out of control.

Attempts to control purple loosestrife by mowing, burning, flooding, and using herbicides have not been very successful. These methods are either too costly, ineffective, or they must be repeated over and over again. Scientists are now trying a method called biological control, which uses the natural enemies of the introduced species such as insects and pathogens (bacteria and viruses). However, this method is quite risky, as it involves introducing yet another species to the area. A great deal of care must be taken before undertaking biological control. In 1992, scientists in New York tested three kinds of introduced insects to attempt to control loosestrife. After careful study, the insects were released and the results look promising. In one study area, purple loosestrife was reduced by 75 to 80 percent, and the insects have caused no secondary damage to the wetlands.

Hydrilla is another invasive plant that causes problems for wetlands in the southeastern United States. Native to Africa, Asia, and Australia, it is a plant used in home aquariums that found its way into wetlands when people dumped their aquariums out. Hydrilla is adaptable to most kinds of water and therefore survives very well in a variety of wetlands. In Florida, officials spend $14.5 million a year in herbicides to keep waterways clear of hydrilla.

Once an invasive species has entered the ecosystem, it is very hard to remove. Effort is being made to prevent the spread of introduced plants and animals by educating the public and inspecting food produce and other imported items. Another way people can help prevent the spread of introduced species is to use only plants native to their area in their yards. They can also make sure to clean boots, boats, and other equipment that can accidentally carry seeds from one place to another. In addition, no one should release aquarium plants or fish into the wild. Another tactic is to learn the native plants in the local area and report new plants to state conservation offices.

Invasive animals

Plants are not the only alien species that can cause problems in a wetland. Non-native animals can also disrupt ecosystems in many of the same ways and for the same reasons that non-native plants do. Some of the most common invasive animals are rats, cats, carp, and a variety of insects.

One mammal that is causing a problem in the southeastern United States is the nutria, a semiaquatic rodent similar to a muskrat. One hundred thousand acres of wetlands in Louisiana alone have been devastated by nutria. Originally from South America, nutria were brought to the United States for use in the fur industry. Through escapes from fur farms and deliberate release to the wild, nutria now live in

Nutria, non-native, muskrat-like animals originally from South America, now wreak havoc on wetland ecosystems in fifteen U.S. states.

fifteen states. It is estimated that Louisiana alone has a population of 20 million nutria. Nutria can begin to breed at six months old and can produce two litters a year. Each litter has about four or five young. By having so many babies, nutria can withstand a lot of predators. Alligators are nutria's only natural predator, but there are too many nutria and too few alligators to consume them all.

The problem for wetlands is that this huge nutria population has to eat, and in foraging through the wetlands, they end up destroying them. Nutria dig up plants and roots, making it difficult for the plants to reproduce. They also overturn peat layers as they dig through the wetlands. When the plants are eaten, the soil held by the plants is washed away, beginning the devastating process of erosion.

In the fall of 2002, the Louisiana government set aside $1.6 million to trap nutria, promising trappers $4 for every nutria they kill. The goal is to trap 400,000 nutria. "We hope if we do that for three to five years we'll see a reversal in damage,"[15] states Greg Linsombe, a wildlife biologist in Louisiana. In Maryland, the plan is to eradicate the nutria completely. Currently, trappers are testing the best method for eliminating them. One plan to cut the population is to create a market demand for nutria as a food source for humans. Nutria meat is high in protein and carbohydrates and lower in fat than most meat. It remains to be seen if the new market for nutria will catch on.

Recreation

In many ways, people themselves are an introduced animal that threatens wetlands. Ironically, while more and more people appreciate the beauty and wonder of wetlands, they also damage them. Increasing numbers of people are visiting wetlands for boating, hiking, riding off-road vehicles, studying nature, and enjoying other recreational activities.

Even when people are well intentioned, they can damage a wetland by their sheer numbers. People who hike, camp,

Tourists feed an alligator during a wetland sightseeing excursion, unknowingly disrupting the animal's innate behavior.

drive, and stay in parks that feature wetlands can have an adverse effect on the wildlife, plants, and ecology of the wetlands by stepping on delicate plants, scaring off wildlife, bringing in introduced species, or accidentally polluting the water. Animals are disturbed by the presence of people. For example, wakes from boats can not only damage shorelines and cause erosion but disrupt birds' nests and other animal homes. In addition, manatees are very slow and can be injured or even killed when hit by motorboats.

Sometimes animals become too accustomed to having people nearby and lose their natural fear of humans. This can lead to dangerous encounters; wetland animals have caused damage to human property and in some cases even physical injuries. One place where this is happening is Kakadu National Park in Australia, where there are large areas of wetlands and an explosion in tourism. As this type of recreation increases, it will take careful planning to reduce the dangerous impacts on the wetlands.

A situation described in the March 24, 2002, issue of the *New York Times* highlights how people enjoying wetlands can also be damaging them. At the Big Cypress National Wildlife Refuge in Florida, off-road vehicles have created thousands of miles of deep ruts that affect the water quality and flow. This has in turn changed animal habitats and plant growth. Noise from the vehicles and people has also frightened endangered species such as the Florida panther and the Cape Sable seaside sparrow.

The opinions of people who drive off-road vehicles and environmentalists often vary widely. According to Erich Pica of Friends of the Earth, a national environmental organization, "What you have is a minority of the population using the Big Cypress and other places around the nation as an amusement park for their thrill vehicles."[16] On the other side of the issue, Lyle McCandless of the Collier County Sportsmen's Club responds, "You can't have an area that is public and have people come in and not leave a footprint."[17] The National Park Service has begun a plan that will limit off-road vehicles to four hundred miles of designated roads over the next ten years. Opinions on these kinds of issues make preserving wetlands a challenge, requiring a balance among various interests.

Another type of conflict between recreation and wetlands exists along many coastlines. People enjoy vacations to the beach, but many wetlands are found along the ocean. The wetlands are often destroyed to create more beaches and to make room for hotels and other buildings. In France, for example, many resorts were built along the southern shore of France. These resorts attract more than 1.5 million visitors each year. The wetlands that once were there, however, are now gone. This situation has happened in many warm-weather vacation destinations around the world.

Today, some of the most serious threats to wetlands are pollution, resource extraction, and invasive species. To ensure that wetlands will continue to function and survive, there are a variety of programs in place to protect them for generations to come.

5

Protecting Wetlands

UP UNTIL THE mid-1930s, the only laws regarding wetlands in the United States encouraged their destruction (for developmental purposes), not their protection. In recent times, the federal government has shifted from advocating the conversion of wetlands for construction to protecting wetlands. However, there is still no single national law regarding wetlands. The protection and management of wetlands falls under many laws intended for other reasons, primarily water quality and land use.

There are three major ways that wetlands are protected. One is through legislation that prevents or limits damage to wetlands by development and pollution. Existing wetlands are also purchased with the intention of being preserved. Finally, some wetlands are restored. All three ways of protecting wetlands involve governmental agencies, nongovernmental groups, and private individuals. For the most part, these efforts have been successful, and since the late 1970s, there has been a decline in wetland loss.

Early legislative efforts

It was not until the 1900s that the federal government began to shift its focus from draining wetlands to protecting wetlands. The stimulus for this shift was the dramatic decline in waterfowl populations. This decline spurred the creation of two programs, the National Wildlife Refuge System and the Duck Stamp Program. These two programs proved to be effective measures that protected millions of acres of wetlands in the United States.

A duck stamp must be purchased in order to legally hunt ducks. Proceeds from the stamps go towards preserving wetlands.

The national wildlife refuges were wetlands primarily set aside to protect migration routes and breeding grounds for waterfowl. The first refuge was set aside by President Theodore Roosevelt in 1903. Today, there are 492 refuges. They range in size from just a few acres to millions of acres. National parks, national forests, and other federal lands are also places where wetlands are preserved and protected.

However, by 1934 there was still a decline in waterfowl, due mostly to lack of funding, so the Duck Stamp Program began as a way to raise money to protect wetlands for waterfowl. That year, President Franklin Roosevelt appointed a committee to determine how to save the waterfowl. The committee discovered it would take about $50 million to buy the land needed to create enough refuges. Their idea to raise the money was to create a duck stamp. Duck stamps are not used for postage; they are stamps that must be purchased by any waterfowl hunter over the age of sixteen. Each year, a contest is held to select the painting that will be presented on the stamp. The proceeds from the stamps go to buy wetlands and preserve the waterfowl's breeding grounds and migratory routes. The program is still in place, and to this date it has generated $501 million for the preservation of wetlands.

Hunters were not the only ones who saw the need to protect wetlands. In the 1960s, the importance of coastal wetlands was revealed. Once people began to understand coastal wetlands' value as breeding grounds and nurseries for commercial and recreational fisheries, support grew for their protection. Wetlands were no longer viewed as simply good places to put a landfill; they were places worthy of protection. In 1962, Massachusetts passed the first law that recognized the benefits of coastal marshes. Other states followed Massachusetts's lead and joined in protecting the coastal wetlands.

The federal government also became more involved in protecting coastal wetlands. Wetlands are given some protection in the Coastal Zone Management Program, which was established in 1972. This program gives money to states if they develop plans for coastal management that include ways for protecting wetlands. Similarly, the National Flood Insurance Program, established in 1973, gives protection and flood insurance to states that take advantage of the flood protection wetlands offer by avoiding development in flood-prone areas.

Increased protection for wetlands

Wetland protection took another step forward in May 1977 when President Jimmy Carter issued two executive orders that made protecting wetlands an official policy of the federal government. (Orders are not laws, but they guide the work of government agencies.) Executive Order 11990 begins with the following:

> Each agency shall provide leadership and shall take action to minimize the destruction, loss or degradation of wetlands, and to preserve and enhance the natural and beneficial values of wetlands in carrying out the agency's responsibilities for (1) acquiring, managing, and disposing of Federal lands and facilities; and (2) providing federally undertaken, financed, or assisted construction and improvement; and (3) conducting Federal activities and programs affecting land use, including but not limited to water and related land resource planning, regulating, and licensing activities.[18]

Restoring Wetlands with Beavers

An interesting restoration project is taking place on the San Pedro River in Arizona. The San Pedro was once a much different river than it is today. The river used to move slowly through wetlands and numerous beaver ponds. The wetlands and beaver ponds spread the river out on both sides and helped to water a wide area in the middle of a desert. However, water diversion for mining, overgrazing, wood cutting, trapping, and other human activity changed the river. Beaver dams were even dynamited to drain the wetlands. Flash floods were not slowed down, and floodwaters rushed through the river channel, further narrowing it and washing away plants.

In 1988, Congress created the San Pedro Conservation Area, covering a forty-mile stretch of river. To help restore the wetlands, beavers are being released into the area. The beavers build dams that slow the water down and spread the river out. In the shallower, slow-moving water, wetlands can become established along the river's edge. The hope is that the beavers will set up about twenty colonies along the forty-mile stretch of river by the year 2005. In doing so, the beavers will also be restoring the important functions provided by the wetlands. The beavers are being closely studied to learn more about their habitat restoration abilities for use in other places.

The San Pedro River and wetlands are coming back to life. Wetland plants have returned, followed by a stunning variety of birds and other animals. The National Audubon Society designated the San Pedro a globally important bird area. Groundwater pumping for the growing population in the area still threatens the San Pedro, but with the return of the beaver there is hope.

Local beavers are largely responsible for reviving Arizona's San Pedro River wetland.

Executive Order 11988 for floodplain management orders the same protection for floodplains and requires government agencies to avoid building on floodplains. Both of these orders meant that federal agencies had to review their policies on wetlands and make protecting wetlands more of a priority.

In 1987, another step was taken to develop a national policy on wetlands. The National Wetlands Policy Forum was formed at the request of the Environmental Protection Agency. The forum developed two main objectives: "To achieve no overall net loss of the nation's remaining wetlands base and to create and restore wetlands, where feasible, to increase the quantity and quality of the nation's wetland resource base."[19] The no-net-loss concept means that when a wetland is destroyed for development, then another wetland must be constructed or restored to take its place.

One of the most important pieces of legislation that is used to achieve these goals and protect wetlands is the Clean Water Act. The Clean Water Act, which is responsible for improving the water quality in the United States, was first passed in 1972. One part of the Clean Water Act, Section 404, has been instrumental in helping to protect wetlands. Section 404 states that anyone dredging or filling in waters in the United States must request a permit from the U.S. Army Corps of Engineers. For example, if a developer wants to build a shopping mall in a wetland, he or she must first get permission from the Army Corps. The Army Corps then takes a look at a variety of environmental and economic factors before deciding whether to give the developer permission to destroy the wetland. There is also an opportunity for the general public to make comments on the proposed projects, which gives local citizens some input on the decision.

Section 404 of the Clean Water Act has protected many wetlands, but only about 20 percent of the activities that threaten wetlands are regulated by the legislation. For example, agricultural land and land used for timber harvests are exempt from Section 404 regulations. Therefore, most wetlands still need other forms of protection.

In 1985, the "swampbuster" provision of the Food Security Act was passed. This provision denies federal money to farmers who knowingly destroy wetlands. The provision helps farmers by giving them assistance in identifying wetlands and developing ways to protect them. Initially, this program was unpopular with most farmers and there was a lack of enforcement. It took a while for farmers to accept a change in thinking from draining wetlands to protecting them. Over time, however, support grew as more farmers viewed wetlands as a resource, not just wasted farmland.

Recent legislative action

Another way that the federal government is helping farmers protect wetlands is the Wetlands Reserve Program (WRP), which began in 1990. The WRP is a voluntary program landowners can use to protect, restore, and enhance

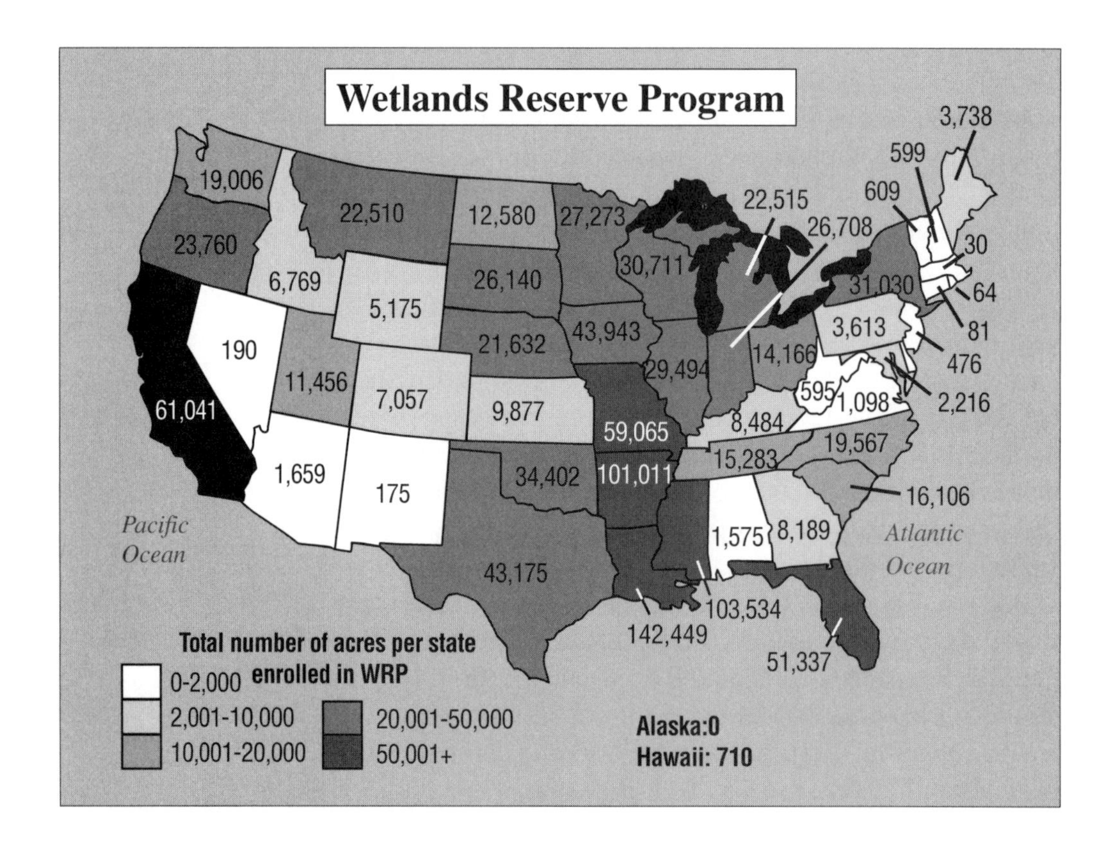

wetlands on their property. WRP offers payment to restore wetlands in places where they have previously been drained and converted to farms. The program pays up to 100 percent of the costs for restoring the wetlands. The landowners retain control of the land, and the land can still be sold.

The wetlands that are eligible for the restoration program are ones that have been cleared or drained. In addition, places near wetlands that provide water and help contribute to the functions of the wetland can also be restored under the program. Since 1999, 318,000 acres of wetlands have been restored, mostly in the Lower Mississippi River Valley, Iowa, and Florida.

The Farm Conservation Bill, which was signed into law on May 14, 2002, continues the trend toward more wetland protection. The main focus of the bill is to support farmers and ranchers. The legislation helps wetlands because farms and ranches represent two-thirds of private land in the United States and 75 percent of all wetlands are on this privately owned land. Since so many wetlands are on these private lands, the support of farmers and ranchers in wetland protection is important. The Farm Conservation Bill includes substantial increases in the money spent on the Wetlands Reserve Program.

This added support for farmers will help preserve wetlands throughout the United States. As Ken Babcock, director of operations for Ducks Unlimited's Southern Regional Office, says, "Simply put, more WRP means more benefits for waterfowl."[20] The benefits of more wetland protection are not limited to waterfowl but to people as well.

International efforts

Since wetlands do not recognize political borders, protecting them is an international concern. In 1971, an intragovernmental treaty was signed in Ramsar, Iran. The Ramsar Convention was the first modern global treaty on conservation. The objective of the treaty is to reduce wetland loss and recognize the functions of wetlands as well as their social and scientific value. Currently, 117 countries have signed the Ramsar Convention, including the United States.

The original focus of the Ramsar treaty was to provide habitat for waterbirds, since waterbirds do not acknowledge international boundaries and need help everywhere along their migration routes. The convention has since broadened its scope to cover all aspects of wetland conservation. The convention designates wetlands of international importance. They range in size from the 16,954,080 acres in Okavango Delta in Botswana to the 2.4 acres of Hosnie's Spring in Australia. Each wetland must meet certain criteria to be designated. Some of the criteria include wetlands that support threatened or endangered animals and plants, wetlands that are important to the life cycles of fish, and wetlands that support waterfowl on an international scale. So far, there are more than 1,070 wetlands of international importance.

Another example of international cooperation is the North American Waterfowl Management Plan. This treaty between the United States and Canada helps protect migrating waterfowl. Through both private and public support, $1.5 billion has been raised to protect wetlands that cross the border between the two countries.

Waterfowl populations, like Pintail ducks, are recovering thanks to environmental advocacy groups and international legislative action.

Nongovernmental efforts to protect wetlands

In addition to federal, state, and local governments, many environmental groups and private citizens are working to protect wetlands. The goal of many of these groups is to preserve habitat for fishing and hunting. One of the major groups is Ducks Unlimited. Ducks Unlimited has 1 million supporters and has raised more than $1.5 billion since 1937 to conserve 10 million acres of wetlands in Canada, the United States, and Mexico. Ducks Unlimited was founded out of concern for the declining duck population after a severe drought hit North America in the 1930s. Founder Joseph P. Knapp originally named the organization More Game Birds in America. The organization started its efforts in Canada, where most waterfowl breed. Over time, it changed its name to Ducks Unlimited and went on to organize more than twenty-six thousand projects to protect waterfowl and other wildlife. Some of the projects involved acquiring land, working with landowners, replanting forested wetlands, restoring wetlands, and conducting scientific research.

Other organizations such as the National Wildlife Federation and the Izaak Walton League have joined the effort to preserve wetlands. Groups led by the National Audubon Society and the Nature Conservancy work to protect wetlands for the sake of the animal and plant habitats they provide. Other organizations such as the Sierra Club and the Wilderness Society have also recognized the need to protect wetlands and to preserve their functions. These groups educate the public, conduct research, preserve land through acquisition, and work to influence legislators.

Opposition to wetland protection

There has been some opposition to various aspects of wetland protection. As the federal government has gained more control over wetlands, some private landowners have taken issue with government interference. They object to the lack of freedom to use their land the way they please. There is also debate over how much money landowners deserve when

A California Conservation Corps member plants a native valley oak in order to restore a wetland to a healthy state.

they are denied the use of their land in order to preserve wetlands. Developers have argued that a small wetland that has few valuable functions is less important than the housing that could be built in that spot. Groups that advocate private-property rights have gained support in some circles and have slowed the pace of wetland protection. Their efforts have been successful in changing the definition of a wetland to include fewer areas. As a result, some wetlands are now developed without public review.

Wetland restoration

Federal and state officials, as well as private organizations, are taking advantage of the fact that wetlands can sometimes be restored. Wetland managers try to restore wetlands back to their previous state. In some cases, new wetlands can even be created.

The first and most important step in restoring a wetland is to reestablish the proper water level in the wetland. This may involve breaking up drainage tiles or plugging a canal

Wine Country Wetland Restoration

Inspired individuals can make a difference when it comes to preserving wetlands. One success story is the Viansa wetlands in Sonoma County, California. Sam Sebastiani, the former president of Sebastiani Vineyards, had a vision for some of the land that he had recently purchased for a winery. It also seemed to him a perfect place to create a marsh. In his article "Just Add Water," Kenneth Brower describes the inspiration that Sam Sebastiani saw that moved him to restore the marsh on his land: "Something in the lay of the land—in the way the flat embayment of the field nested in an amphitheater of hills—must have murmured marsh to him; for now he saw a kind of mirage. The field flooded in his imagination, its stubble greening and lengthening into stands of cattails and the 20-foot culms [stems] of bulrushes, with mallards and pintails swimming underneath them."

Upon further investigation, Sebastiani learned that the land he was looking at had been a wetland up until the time of the California gold rush in 1849. All the land needed was for the water to return. To re-create the marsh, Sebastiani built a seventeen-hundred-foot levee to catch the water from winter rains and prevent it from draining away. After the first winter, the water stayed, and dormant seeds from the old wetland grew into plants. Other seeds came in with the wind and more wetland plants returned. With the return of the water and plants, animals began visiting the wetland. On a single winter day, ten thousand waterfowl were counted on its ninety acres. Nearly every North American duck species has since paid a visit. Other species of birds come to the marsh, too, from hawks to sparrows. The success of the Viansa marsh proves that some wetlands can be restored with the right kind of care and attention.

With help from a new levee, the Viansa Wetland Preserve now stretches across ninety acres of California wine country.

that has taken the water away. If a wetland has been dredged out, fill must be put back in to make the water shallow enough to support wetland plants. Once the water levels are restored, then the wetland plants can return. The plants can sometimes return on their own, but many are replanted by the people restoring the wetland. Once the plants are growing again in the wetland, animals can return, and the functions of the wetland can begin again.

Wetland managers who work to restore and create wetlands have many things to keep in mind to make the project work. A plan with minimum maintenance will not only keep the financial costs down but reduce the amount of direct human involvement needed. Care must be taken to select a suitable site for restoring or creating wetlands. Wetlands restoration is easier than wetlands creation, so it makes sense to try to use areas that have been or still are wetlands. Wetlands need space, so it is important to select a site that has enough land and to consider the surrounding area. A road or housing development that may end up being built near the new wetland, for example, could hamper the project. With planning and effort, wetlands can be created and restored, but it is not easy.

Evaluating the success of wetland restoration

Numerous research attempts have been made to measure the success of wetland restoration. This is the only way of figuring out which methods work and which do not. Although there are examples of successful projects, the studies show that there is a need to improve the methods used for wetland restoration.

A report by the National Research Council (NRC) of the National Academy of Science found that the no-net-loss policy is not being met. From 1993 to 2000, some twenty-four thousand acres of wetlands were filled, and forty-two thousand acres were to be restored to replace the ones that were lost. That looks like a 74 percent gain of wetlands. However, this is not what happened. Many of the planned restoration projects never began. In Indiana, for example, only 62 percent of the proposed wetlands were created.

Other research supports the findings of the NRC. One study of wetland restoration projects in Florida found that only half of the proposed wetlands are being constructed. In another study of an area around Chicago, 128 sites were studied. In only 17 sites did the planned vegetation become established.

One reason for the problem is that the Army Corps is not doing an adequate job of keeping track of wetland restoration projects or enforcing the regulations. Another problem is that, in many wetland projects, there is a lack of understanding on how to re-create a wetland. "We simply don't know how to restore or construct all types of wetlands," says Joy Zedler, chair of the NRC study. "Re-creating the complex chemistry of bogs and fens is beyond us."[21] This is compounded by the fact that many of the people constructing the wetlands lack the proper training or are simply trying to build the wetland as cheaply as possible.

Restoring the Everglades

A major restoration project is taking place in the Florida Everglades. Known as the River of Grass, the Everglades are a unique subtropical wetland originally covering 1.5 million acres. The Everglades are home to an astounding amount of wildlife and are vital to the water supply of south Florida. However, because of human actions over the past one hundred years, the Everglades are in trouble.

The Everglades are dependent on the water from Lake Okeechobee, located north of the Everglades, as well as rainfall. Each spring, water overflows the lake's banks and slowly flows south in a fifty-mile-wide sheet of water that is only inches deep. However, most of this water is no longer reaching the Everglades. The water is diverted away from the Everglades for agricultural uses, drinking water, and flood-control measures. The water that does reach the Everglades comes during the wrong seasons and is not nearly enough. In addition, about half of the Everglades have been developed for human uses. As a result, there is severe habitat fragmentation. Runoff from agriculture has also put excess nutrients in

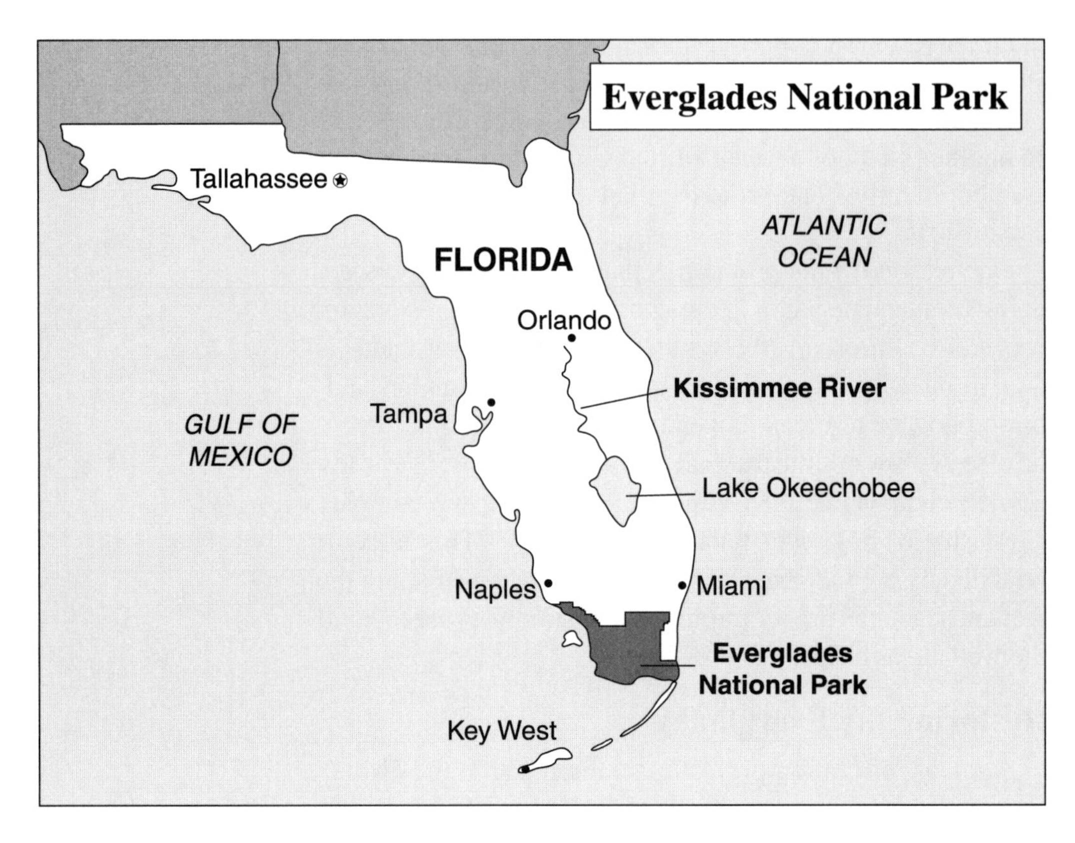

the water, causing the spread of cattails over the native saw grass. One indication of the problems facing the wetland is that 90 percent of the bird population is gone.

Restoring the Everglades is an enormous project covering more than a million acres. The plan is being directed by the Army Corps with help from other federal, state, and local government agencies. The restoration, begun in 2000, will cost $8 billion and is a thirty-year plan.

One of the major goals of the plan is to restore the water in the Everglades. According to the plan, 80 percent of the water that now goes out to sea will be redirected back to the Everglades. This is accomplished through water preserve areas, above-ground reservoirs that will catch the water that once flowed into the ocean. This water will now be used for human consumption and for the Everglades. Additional water will come from urban runoff, which will be filtered by treatment wetlands to lessen the nutrient

Champion for Wetlands

If it were not for Marjory Stoneman Douglas, there might not be much of the Everglades left to restore. In the 1920s, as an assistant editor for the *Miami Herald,* Douglas became involved in the fight to create a national park in the Everglades. Her newspaper columns became a powerful voice in creating support for the park. It took twenty-five years, but in 1947 the efforts of Marjory Stoneman Douglas and many others paid off when President Harry Truman officially opened the Everglades National Park.

Douglas's efforts to protect the Everglades did not end there. When plans were made in 1968 to build a jetport in the Everglades, Douglas, at age seventy-eight, began an organization called Friends of the Everglades. To gain support, she started making speeches to every organization that would listen. She was able to recruit fifteen or twenty new members, at $1 apiece, every time she spoke. The jetport project was stopped, and Friends of the Everglades continued. The focus eventually shifted to the water problems facing the Everglades.

Even after her one hundredth birthday, when she lost her sight, Douglas continued to fight for the Everglades. In 1993, at the age of 103, she was awarded the Medal of Freedom, the highest presidential honor given to a citizen. Marjory Stoneman Douglas died at 108, a true friend of the Everglades.

Marjory Stoneman Douglas, champion of the Everglades National Park, examines the local flora.

load. These special wetlands are called stormwater treatment areas. In addition, levees, dams, and canals will be removed, allowing more of the water to return to its natural flow. A huge part of the plan is to restore the Kissimmee River. This river once wound its way through a ninety-six-mile course. For flood-control purposes, the river was reconstructed into a straight line, and 65 percent of the wetlands along the river were destroyed. With the river restored to its natural design, the Everglades will benefit from decreases in nutrients and an increase in water.

Another part of the restoration plan involves land acquisition. Attempts are being made to buy undeveloped parts of the Everglades and preserve them. The problem is that the state of Florida has less money to spend than developers, who are driving the price of land higher. South Florida's population has grown to 6 million, and it is expected to double in the next fifty years.

One aspect of the plan that is going well is the Southern Golden Gate Estates Project. This is a combination of land acquisition and water-quality improvement projects. The project is reconnecting Big Cypress Swamp to the Everglades, which will improve habitat for the endangered Florida panther and other wildlife. Many people are optimistic that the massive Everglades massive restoration project will succeed. There is hope for the future.

The values and benefits of wetlands are becoming well known. And scientists are continuing to learn more about the role wetlands play in local and global ecosystems. Although efforts are being made to protect and restore wetlands, they still face a variety of threats. It will require continued efforts today and in the future to protect wetlands so that wetlands can protect us and future generations.

Notes

Chapter 1: What Is a Wetland?

1. Quoted in William J. Mitsch and James G. Gosselink, *Wetlands.* New York: John Wiley & Sons, 2000, p. 29.
2. Steve Rice, telephone interview with the author, December 2001.
3. Matthew Draud, Terrapin Conservation Project pamphlet, C.W. Post University, 2001.

Chapter 2: The Benefits of Wetlands

4. Jon Nordheimer, "Nothing's Easy for New Orleans Flood Control," *New York Times,* April 30, 2002, p. F1.
5. Dan Ramer, interview with the author, Bayville, NY, May 18, 2002.
6. Don Stap, "Living on the Edge," *Audubon,* March/April 2002, p. 58.
7. Steve Rice, interview with the author, Albany, NY, October 12, 2002.

Chapter 3: Wetland Destruction

8. John Turner, interview with the author, Cold Spring Harbor, NY, August 10, 2002.
9. Quoted in Andrew C. Revkin, "U.S. Sees Problems in Climate Change," *New York Times,* June 3, 2002, p. 1.

Chapter 4: Wetland Degradation

10. Quoted in *New York Times,* "Bush to Allow Mining Industry to Fill Streams," May 4, 2002, p. 10.
11. *New York Times,* "Burying Valleys, Poisoning Streams," May 4, 2002, p. A12.
12. Mike Matheson, interview with the author, Boulder, UT, April 10, 2002.

13. Quoted in Alan Hall, "Costly Interlopers," *Scientific American,* February 15, 1999. www.sciam.com.
14. Dan Carroll, "Purple Loosestrife: Controlling a Colorful Nuisance," *New York State Conservationist,* August 2000, p. 12.
15. Quoted in Sam Smith, "Fur Trappers Are Taking On the Scourge of the Marshlands," *New York Times,* May 28, 2002. p. F4.
16. Quoted in Dana Canedy, "Land Advocates and Drivers Reach Fork in the Off-Road," *New York Times,* March 24, 2002, p. 33.
17. Quoted in Canedy, "Land Advocates and Drivers Reach Fork in the Off-Road," p. 33.

Chapter 5: Protecting Wetlands

18. Quoted in Mitsch and Gosselink, *Wetlands,* p. 640.
19. Quoted in Mitsch and Gosselink, *Wetlands,* p. 642.
20. Quoted in *Ducks Unlimited,* "Ducks Unlimited and Rice Farmers Work Through Differences to Support Favorable Farm Bill," May 24, 2002. www.ducks.org.
21. Quoted in Erik Ness, "Murky Deal," *Audubon,* November/December 2001, p. 13.

Glossary

acid deposition: Also known as acid rain, precipitation that has a PH lower than normal. It is created by the combination of water molecules and sulfur dioxides and nitrous oxides.

aquifer: An underground bed of soil or rock that is saturated with water.

brackish: Water that is a mix of salt water and freshwater.

decomposer: An animal, fungus, or bacteria that feeds on dead plants and animals and releases the stored nutrients to the soil.

detritus: Dead bits of plants and animals that fall to and settle on the bottom of a wetland.

diversity: In terms of an ecosystem, the number of different plant and animal species found in a given area.

ecosystem: A community of living things and their relationships with the environment.

eutrophication: The change in an aquatic ecosystem from having low levels of nutrients to high levels of nutrients.

floodplain: Land that is usually dry but can be flooded by water from any natural source. Floodplains are often found along rivers and lakes.

global warming: The rise in average temperatures around the world caused by increased amounts of greenhouse gases in the atmosphere.

habitat: The place where an animal or plant lives.

habitat fragmentation: When a habitat is broken up into smaller units by development or other causes, resulting in a number of isolated smaller habitats.

hydric soil: Soil that is full of water for a long enough period over the years to become oxygen depleted on the top layers.

hydrophyte: A plant that is adapted to living in wet conditions.

levee: An embankment or raised area that prevents water from moving from one place to another.

respiration: The process by which plants and animals supply their cells with oxygen and release carbon dioxide.

sediment: Small bits of inorganic and organic material removed from soil and rocks by erosion and weathering.

subsidence: The sinking of the ground level caused by the natural and artificial settling of sediment.

Organizations to Contact

Ducks Unlimited
One Waterfowl Way
Memphis, TN 38120
(800) 45-DUCKS
www.ducks.org

Ducks Unlimited is a leader in wetland and waterfowl conservation. DU has helped to conserve 10 million acres of wetlands in North America.

Environmental Concern, Inc.
PO Box P
201 Boundary Lane St.
Michaels, MD 21663
(410) 745-9620
www.wetland.org

Environmental Concern, Inc., is dedicated to informing educators about wetlands, with two educator guides and workshops.

National Audubon Society
700 Broadway
New York, NY 10003
(212) 979-3000
www.audubon.org

The Audubon Society began as a bird conservation organization but has grown to be one of the largest and most effective environmental protection groups in the United States. It is very involved in the Everglades restoration project.

National Wildlife Federation
11100 Wildlife Center Dr.
Reston, VA 20190
(703) 438-6000
www.nwf.org

The National Wildlife Federation is the largest conservation group in the United States. It works to protect the wonders of nature at home, in schools, and across the country.

Nature Conservancy
4245 North Fairfax Dr., Suite 100
Arlington, VA 22203
(800) 628-6860
comment@tnc.org

The motto of the Nature Conservancy is "Saving the Last Great Places." The Nature Conservancy works with private landowners and local governments to purchase natural areas for protection.

Suggestions for Further Reading

Tricia Andryszewski, *Marjory Stoneman Douglas: Friend of the Everglades.* Brookfield, CT: Millbrook Press, 1994. A biography of Marjory Stoneman Douglas, the woman who used her influence as a writer and environmentalist to protect the Everglades for more than sixty years.

Paul Fleisher, *Salt Marsh.* New York: Benchmark Books, 1999. A good introduction to the interrelationships of a salt marsh. The photographs will help the reader get a good picture of the animals and plants that live in a salt marsh.

Elizabeth P. Lawlor, *Discover Nature in the Water and Wetlands: Things to Know and Things to Do.* Harrisburg, PA: Stackpole Books, 2001. A great book with information on all aspects of wetlands as well as activities for experiencing wetlands hands-on.

William A. Niering, *Wetlands.* New York: Alfred A. Knopf, 1985. A field guide to the animals and plants of North American wetlands.

Donald Silver, *One Small Square: Swamp.* New York: W.H. Freeman, 1997. Focusing on an exploration of one square meter of a swamp, this book includes activities and information on life in a swamp.

Frank Straub, *America's Wetlands.* Minneapolis, MN: Carolrhoda Books, 1995. A good overall introduction to the ecology of wetlands and the efforts to protect them.

Dave Taylor, *Endangered Wetland Animals.* New York: Crabtree, 1992. This book describes the lives of nine endangered animals that live in wetlands. There is also some information on the threats to wetlands and wetland preservation.

Works Consulted

Books

William H. Amos and Stephen H. Amos, *The Audubon Society Nature Guides: Atlantic and Gulf Coasts.* New York: Alfred A. Knopf, 1985. This is primarily a field guide to animals and plants of the Atlantic coast, but it includes a good introduction to the ecology of salt marshes and mangrove swamps.

Michael J. Caduto, *Pond and Brook.* Hanover, NH: University Press of New England, 1990. A comprehensive book covering the ecology of a variety of aquatic habitats, including wetlands.

Charles W. Johnson, *The Nature of Vermont.* Hanover, NH: University Press of New England, 1998. Though the focus is on the natural history of Vermont, the information on the various wetlands found there applies to other places as well.

William J. Mitsch and James G. Gosselink, *Wetlands.* New York: John Wiley & Sons, 2000. An extremely comprehensive textbook covering just about every aspect of wetlands.

Bill Streever, *Bringing Back the Wetlands.* Potts Point, Australia: A Sainty Book, 1999. The story of a wetland scientist's efforts to restore wetlands in Australia.

John and Mildred Teal, *Life and Death of the Salt Marsh.* New York: Ballantine Books, 1969. A classic book that describes the origins of a salt marsh, life in a salt marsh, and the destruction of salt marshes by human development.

Ralph W. Tiner, *In Search of Swampland.* New Brunswick, NJ: Rutgers University Press, 1999. A comprehensive look at the ecology of wetlands, threats to their existence, and how to protect them. The book focuses on wetlands of the Northeast and includes an identification guide.

John L. Turner, *Exploring the Other Island: A Seasonal Guide to Nature on Long Island.* Great Falls, VA: Waterline Books, 1994. Turner's book highlights interesting natural phenomena all over Long Island, season by season.

Ann Vileisis, *Discovering the Unknown Landscape: A History of America's Wetlands.* Washington, DC: Island Press, 1997. A detailed and interesting account of the relationship between wetlands and humans in the United States.

Michael Williams, ed., *Wetlands: A Threatened Landscape,* Oxford, England: Blackwell, 1990. A series of essays covering a variety of wetland topics. A good source for getting an international perspective on wetland issues.

Periodicals

Leslie Aylsworth, "Hostile Takeover," *Adirondack Life,* July/August 1999.

Kenneth Brower, "Just Add Water: Resurrection of a California Marsh," *Sierra,* November/December 2001.

Dana Canedy, "Land Advocates and Drivers Reach Fork in the Off-Road," *New York Times,* March 24, 2002.

Dan Carroll, "Purple Loosestrife: Controlling a Colorful Nuisance," *New York State Conservationist,* August 2000.

Gretchen C. Daily and Katherine Ellison, "The New Economy of Nature," *Orion,* Spring 2002.

Tom Dollar, "Leave It to Beavers," *Wildlife Conservation,* May/June 2002.

Erik Ness, "Murky Deal," *Audubon,* November/December 2001.

New York Times, "Burying Valleys, Poisoning Streams," May 4, 2002.

———, "Bush to Allow Mining Industry to Fill Streams," May 4, 2002.

Jon Nordheimer, "Nothing's Easy for New Orleans Flood Control," *New York Times,* April 30, 2002.

Andrew C. Revkin, "U.S. Sees Problems in Climate Change," *New York Times,* June 3, 2002.

Sam Smith, "Fur Trappers Are Taking On the Scourge of the Marshlands," *New York Times,* May 28, 2002.

Don Stap, "Living on the Edge," *Audubon,* March/April 2002.

Internet Sources

Arcata Marsh Wildlife Sanctuary, "Saltwater Vegetation," October 10, 2001. www.humboldt.edu.

Discover Magazine, "Pesky Foreign Life-Forms Cost Us Billions," May 1999. www.discover.com.

Ducks Unlimited, "Ducks Unlimited: A Leader in Wetland Conservation," July 3, 2002. www.ducks.org.

———, "Ducks Unlimited and Rice Farmers Work Through Differences to Support Favorable Farm Bill," May 24, 2002. www.ducks.org.

Environment Waikato, "Threats to Wetlands," March 8, 2002. www.ew.govt.nz.

Fish and Wildlife Service, "History of the Federal Duck Stamp Program," September 29, 2002. http://duckstamps.fws.gov.

Alan Hall, "Costly Interlopers," *Scientific American,* February 15, 1999. www.sciam.com.

Laura Helmuth, "Can This Swamp Be Saved?" *ScienceNews Online,* April 17, 1999. www.sciencenews.org.

"Mangroves: More than Mud and Mozzies," November 11, 2001. www.env.qld.gov.au.

National Resources Conservation Service, "Noxious, Invasive, and Alien Plant Species," March 8, 2002. www.pwrc.usgs.gov.

National Wetlands Research Center, "Coastal Wetlands and Global Change: Overview," July 18, 2002. www.nwrc.gov.

———, "Nutria: Eating the Louisiana Coast," April 20, 2001. www.nwrc.usgs.gov.

William K. Stevens, "Putting Things Right in the Everglades," *New York Times*, April 13, 1999. www.nytimes.com.

"Values and Functions of Wetlands," October 10, 2001. www.epa.gov.

Watersheds, "Functions of Wetlands," March 3, 2002. http://h2osparc.wq.ncsu.edu/info.

"Wetland Soil," November 11, 2001. www.wetland.org/kids/wetsoil.htm.

"What Is a Wetland?" November 18, 2001. www.wetland.org/kids/wetlands.htm.

"What Wetlands Do for You," March 1, 2002. www.wetland.org/kids/foryou.htm.

"Whooping Cranes," May 7, 2002. www.ngpc.state.ne.us.

Websites

Environmental Protection Agency (www.epa.gov). This site has fact sheets on all aspects of wetlands, from ecology to regulations.

U.S. Fish and Wildlife Service National Wetlands Inventory (www.nwi.fws.gov). This site has information on the national refuges system and the wetlands inventory.

Index

Africa, 68
agriculture, 8
animal grazing and, 64–65
draining wetlands for, 47–48
Everglades and, 85–86
Farm Conservation Bill and, 79
filling in of wetlands for, 46–47
Food Security Act and, 78
loss of wetlands to, 44
Alaska
bird migration from, 37
glaciers and, 6
wetlands in, 43
Algonquin (tribe), 19
Anhinga Trail (Florida), 40
animals
abundance of, 34
alligators, 69
beavers, 76
bog lemming, 15
Cape Sable seaside sparrow, 71
capybara, 21
dams and, 53
diamondback terrapin, 20
ducks, 40
effect of invasive plants on, 66
in the Everglades, 86
fish, 38
human recreation and, 71
invasive, 68–69
ivory-billed woodpecker, 63
jabiru, 21
loss of habitat and, 44–45
migration routes and, 26, 37
mosquitoes, 47, 50–51
osprey, 60
of prairie potholes, 18
of salt marshes, 20
selenium poisoning and, 65
of swamps, 20
tiger salamander, 23
variety of, 35, 37
vernal pools and, 23, 49–50
whooping cranes, 36
Antarctica, 24
archaeology, 41
Asia, 39, 68
Australia, 19, 68

Babcock, Ken
on waterfowl and WRP, 79
Baptist, William, 63

Baton Rouge, Louisiana, 58
Bay of Fundy (Nova Scotia), 37
Big Cypress National Wildlife Refuge (Florida), 71
Big Cypress Swamp (Florida), 88
Black Swamp (Ohio), 61
Blaxton, William, 45
bogs (peatlands), 6
  animal life in, 15
  creation of, 13
  mummies in, 16
  nutrients for plants in, 14–15
Boston, Massachusetts, 47
Boston Red Sox, 45
Brazil, 21
Brower, Kenneth
  on Sebastiani restoration of wetlands, 83
Bush, George W.
  global warming and, 56
  mining and, 62

California, 17, 61
Canada
  bogs and, 13, 19
  Ducks Unlimited and, 91
  glaciers and, 6
  peat moss and, 39
  prairie potholes and, 18
  waterfowl treaty with U.S., 80
Carrol, Dan
  on abundance of purple loosestrife, 67
Carter, Jimmy
  executive orders by, protect wetlands, 75
Catskill Mountains, 32
cattails, 7
  dams and, 17, 53
  in the Everglades, 86
  uses for, 18
Central Park (New York City), 47
Central Valley (California), 65
Charles River, 27
Civil War, 63
Clean Water Act, 77
climate
  bogs and 13
  flooding and, 26–29
  fens and, 15
  global warming and, 54–55
  prairie potholes and, 18
  salt marshes and 18–19
  storms and droughts 7, 34, 46
  surface water and, 10
Coastal Zone Management Program, 75
Collier County Sportsmen's Club, 71
Cornell University, 66
Corven, Jim
  on sandpipers and wetlands, 37
Crafa, Robert
  on educational value of salt marsh, 31
cypress trees, 10, 39

bald, 22
logging of, 63

Denmark, 19
development
animals and, 45
coastal wetlands and, 48, 72
Everglades and, 85, 88
negative effect of, 47
as opposition to wetland protection, 82
railroads and, 61
Devil's Swamp (Louisiana), 58
diseases
West Nile virus, 50
caused by wetland loss, 44–45
medicines from, 41
Douglas, Marjory Stoneman, 87
Duck Stamp Program, 73–74
Ducks Unlimited, 79
conservation efforts of, 81

economy
cattle grazing and, 64
disaster prevention and, 29–30
fishing and, 38
flood control and, 27–28
railroad and, 61
water processing and, 32, 34
Egyptians, 6
Emergency Wetland Resources Act of 1986, 43
endangered species, 36
Florida panther, 37, 71, 88
Endangered Species Act (ESA), 38
Environmental Protection Agency (EPA), 77
Europe, 19
Everglades National Park
establishment of, 87
tourism in, 40
Executive Order 11988, 77
Executive Order 11990, 75

Farm Conservation Bill, 79
Fenway Park (Boston), 45
Fish and Wildlife Service, 43
flood control, 28–29, 46
dredging and, 52
Everglades and, 85
isolation of wetlands and, 51
wetland function in, 27
Flood Insurance Program, 75
Florida
hydrilla problems in, 68
restoration of wetlands in, 79
timber in, 39
Florida Everglades, 21
fires in, 55
geological formations and, 6–7
plans for restoration, 85–88
plants in, 17
see also Everglades National Park

Food Security Act, 78
freshwater marshes, 17
  animals of, 18
Friends of the Earth, 71
Friends of the Everglades, 87

Galveston Bay, Texas, 54
Geneva, New York, 49
Georgia, 39
Gran Pantanal (South America), 21
Great Britain, 19
  loss of wetlands in, 43

Hosnie's Spring (Australia)
  Ramsar Convention and, 80
Houston, Texas, 54
Hudson River, 32

Illinois, 48–49
Indiana, 49
Intercoastal Waterway, 52
Iowa
  destruction of wetlands in, 48–49
  restoration of wetlands in, 79
Izaak Walton League, 81

Johnston, John, 49
"Just Add Water," 83

Kakadu National Park (Australia), 71
Kalahari Desert (Botswana), 21
Kesterson National Wildlife Refuge (California), 65
Kissimmee River, 88
Knapp, Joseph P., 81

Lake Maurepas (Louisiana), 52
Lake Okeechobee (Florida), 85
Lake Pontchartrain (Louisiana), 28, 52
legislation
  creation of protection programs, 73
  no-net-loss concept, 77
Lewisburg, West Virginia, 62
Linsombe, Greg
  on eradication of nutria, 69–70
"Living on the Edge," 37
Louisiana
  nutria and, 68–69
  subsidence and, 54
  timber in, 39
Lovett, Joe
  on mining and wetlands, 62
lumbering
  clear-cut and, 64
  draining of wetlands for, 63

Maine, 39
mangrove swamps, 23–24
Maryland, 69
Massachusetts
  cranberries in, 39

legislation on coastal wetlands and, 75
Matheson, Mike
on effect of cattle grazing on wetlands, 64
McCandless, Lyle
on effect of humans on wetlands, 71
Medal of Freedom, 98
Mesopotamia, 6
Mexico, 81
Miami Herald (newspaper)
on Marjory Douglas, 87
mining
oil and gas extraction in, 54
peat, 61–62
tailings and, 62
Minnesota, 48
Mississippi, 39
Mississippi River
flooding of, 46
New Orleans and, 28
restoration of wetlands along, 79
More Game Birds in America. *See* Ducks Unlimited

National Academy of Science, 84
National Audubon Society, 76, 81
National Park Service, 71–72
National Research Council (NRC), 84
National Wetlands Policy Forum, 77
National Wildlife Federation, 81
National Wildlife Refuge System, 73–74
Native Americans, 18
Native Utah Cutthroat Association, 64
natural resources
fossil fuels and, 6
timber, 39
wetland forests, 29
Nature Conservancy, 49, 81
estimate of endangered species and nonnative plants, 66
New Jersey, 39
dams in, 52–53
New Orleans, Louisiana
flood control and, 28
pull boat invented in, 63
New York City
built on wetlands, 47
drinking water and, 32, 34
New York State, 29
New York State Conservationist (magazine), 67
New York Times (newspaper)
on coastal marshes and New Orleans, 28
on destruction of wetlands by recreation, 71
on waste dumping, 62
New Zealand
loss of wetlands in, 43
name for cattail swamps in, 19

North American Waterfowl Management Plan, 80
Northern Hemisphere, 13
North Pole, 54

Okavango Delta (Africa), 21
  Ramsar Convention and, 80
Okavango River, 21
Oneida, New York, 32
Oregon, 17
Oyster Bay, New York, 31

Panama Canal, 61
Paraguay, 21
Philadelphia, Pennsylvania, 47
Pica, Erich
  on use of ATVs at Big Cypress, 71
Pimentel, David
  on nonnative plants, 66
plants
  emergents, 12–13
  food for animals, 34–35
  halophytes, 20
  hydrophytes, 11, 17
  insect eating, 15
  invasive, 65–68
  peat moss, 39
  restoration of, 84
  of salt marshes, 20
  sawgrass, 6
  special adaptations of, 11–12
  sphagnum moss, 14
  submergents, 12–13
  swamps and, 20
  trees, 17, 39
  as wetland indicator, 9
Pocono Mountains, 53
pollution
  acid rain, 58–59
  DDT (dichlorodiphenyl-trichloroethane), 59
  excess nutrients and, 60
  filtering of, 57
  global warming and, 42
  from mining, 5, 62
  oxygen cycle and, 41–42
  pesticides and, 48
  types of, 57–58
  water quality, 30–32
prairie potholes (seasonal marshes), 18
Puritans, 45

Ramer, Dan
  on use of wetlands in water treatment, 32
Ramsar, Iran, 79
Ramsar Convention, 79–80
recreation
  destruction of wetlands and, 70–72
  effect on animals 40–41
restoration
  vs. creation, 84
  establishment of proper water level and, 82–83
  use of beavers in, 76
Rice, Steve
  on global warming, 42
  on sphagnum, 14
Roosevelt, Franklin, 74
Roosevelt, Theodore, 74

salt marshes
  biodiversity of, 19
  formation of, 18
San Pedro Conservation Area (Arizona), 76
San Pedro River, 76
Sebastiani, Sam
  restoration of wetland by, 83
Sebastiani Vineyards, 83
Shawmut Peninsula (Massachusetts), 45
Siberia (Russia)
  bogs and, 13
  peat moss from, 39
Sierra Club, 81
soil
  hydric, 11
  as wetland indicator, 6
Sonoma County, California, 83
South America, 68
Southern Golden Gate Estates Project, 88
South Pole, 54
Stap, Don
  on sandpipers and wetlands, 37
swamps, 6
  formation of, 20, 22
  mangrove, 23–25
  vernal pools, 23
Sweden, 19

Texas, 54
Tierra del Fuego (South America), 37
Tollund, Denmark, 16
Tollund Man, 16
Truman, Harry, 87
Turner, John
  on loss of vernal pools, 49

Union College, 14, 42
United Nations, 56
United States
  bogs and, 13
  decreasing loss of wetlands in, 43
  destruction of wetlands in, 73
  Ducks Unlimited and, 81
  endangered species in, 37
  invasive animals in, 68–69
  invasive plants in, 68
  names for wetlands in, 19
  prairie potholes and, 18
  productive plants in, 39
  waterfowl treaty with Canada, 80
U.S. Army Corps of Engineers, 27–28, 77
  inadequate job of enforcement, 85
  restoration of the Everglades and, 86
U.S. Fish and Wildlife Service, 9

vernal pools, 23
  loss to development and, 49–50
Viansa wetlands (California), 83

Waring, George, 47
Washington, D.C., 47

Waterfront Center, 31
Western Hemisphere, 21
Western Hemisphere Shorebird Reserve Network, 37
wetlands
  biodiversity of, 7
  consequences of loss of, 44–46
  creation of, 62
  dams and, 52–54
  degradation of, 57–72
  delicate balance of, 46
  draining of, 47–51
  dredging, 51–52
  educational value of, 40–41
  extent of, 24–25
  fens, 15
  fires and, 55
  food production and, 9, 38–39
  forests in, 39
  groundwater recharge and, 34
  human benefits of, 26
  most abundant, 20
  natural forces and, 6–7
  protection of, 73–88
  research and, 41
  rising sea levels and, 54
  threats to, 7–8
  three indicators of, 9
  types of, 13
  underground minerals and, 54
  value of, 6, 42
  various names for, 19
  as wastelands, 6, 46–47
  water and, 9–11, 24, 51
Wetlands Reserve Program (WRP), 78–79
Whartenby, Benjamin, 49
Wilderness Society, 81
Wisconsin, 39
World War II, 48

Zedler, Joy
  on difficulty of wetland restoration, 85

Picture Credits

Cover Photo: © Cindy Kassab/CORBIS
Ronald L. Bell/U.S. Fish and Wildlife Service, 35
© Jonathan Blair/CORBIS, 58
Tupper Ansel Blake/U.S. Fish and Wildlife Service, 17
© Jan Butchofsky-Houser/CORBIS, 83
© Suzanne L. Collins/Photo Researchers, Inc., 33
© Michael Durham, 7, 10, 12, 76
© Kevin Fleming/CORBIS, 87
© Raymond Gehman/CORBIS, 64
George Gentry/U.S. Fish and Wildlife Service, 38
Luther Goldman/U.S. Fish and Wildlife Service, 36
© Philip Gould/CORBIS, 32, 70
© Hellio/Van Ingen/Photo Researchers, Inc., 60
John and Karen Hollingsworth/U.S. Fish and Wildlife Service, 69
© Chris Lisle/CORBIS, 16
Jeff Di Matteo, 29
Wyman Meinzer/U.S. Fish and Wildlife Service, 80
Brandy Noon, 14, 78
© Reuters NewMedia Inc./CORBIS, 55
© Lynda Richardson/CORBIS, 24
© David M. Schleser/Photo Researchers, Inc., 50
© Blair Seitz/Photo Researchers, Inc., 67
Charles Shuman/U.S. Fish and Wildlife Service, 40
© Joseph Sohm; ChromoSohm Inc./CORBIS, 28
© Sygma/CORBIS, 21
R. Town/U.S. Fish and Wildlife Service, 53
Ned Trovillion/U.S. Fish and Wildlife Service, 22
U.S. Fish and Wildlife Service, 48, 74

About the Author

Dan Kriesberg has worked as a naturalist and an elementary school teacher at the Locust Valley Elementary School in Locust Valley, New York. Currently, Dan has his own environmental education consulting business called Salamander Solutions. He works with schools, nature centers, and other groups training teachers, developing programs, and teaching children. He is the author of *A Sense of Place: Teaching Children About the Environment with Picture Books* and has written articles for numerous magazines.

Dan and his wife, Karen, a second-grade teacher, have two children, Zachary Charles and Scott Walden. They live in Bayville, Long Island, with a salt marsh just down the street.